Coastal and Estuarine Studies

Managing Editors:
Malcolm J. Bowman   Richard T. Barber
Christopher N.K. Mooers   John A. Raven

Coastal
and Estuarine Studies

44

David G. Aubrey and
Graham S. Giese (Eds.)

Formation and Evolution
of Multiple Tidal Inlets

American Geophysical Union
Washington, DC

Editors

David G. Aubrey
Department of Geology and Geophysics
Woods Hole Oceanographic Institution
Woods Hole, MA 02543

Graham S. Giese
Department of Geology and Geophysics
Woods Hole Oceanographic Institution
Woods Hole, MA 02543

Library of Congress Cataloging-in-Publication Data

Formation and evolution of multiple tidal inlets / David G. Aubrey and
   Graham S. Giese (eds.).
      p.   cm. — (Coastal and estuarine studies ; 44)
     ISBN 0-87590-258-8
     1. Inlets.   2. Coast changes.   I. Aubrey, David G.   II. Giese,
Graham S. III. Series.
GB454.I54F67 1993
551.4'57—dc20
                                                    93-38051
                                                       CIP

ISSN 0733-9569
ISBN 0-87590-258-8

CONTENTS

PREFACE

Coastal inlets dot the world's coastline, serving as important conduits for exchange of organic and inorganic materials between the continents and the oceans. In addition, inlets are the focal points of navigation pathways between deep water and coastal ports, a function often requiring alteration of the inlet to assure safe navigation. Particularly for inlets along mobile barrier beaches, frequent dredging is often required to maintain navigation clearances.

The papers in this book primarily discuss research relating to tidal inlets along barrier beach systems and illustrate the scientific approaches appropriate for examining multiple inlet stability. The aim is to provide a scientific basis for effective management. By proper and in-depth scientific analysis, more appropriate inlet management can be effected, to the benefit of the environment and the inhabitants of coastal areas. Case examples are included that will be useful to those who may wish to pursue the wider management and policy implications.

As a result of increasing anthropogenic pressure, natural processes within estuary-lagoon-beach systems are frequently being altered. For example, construction along barrier beaches has altered the response of beaches to coastal storms, dredging of inlets has altered circulation in embayments as well as sediment transport pathways, and jetties protecting inlets have altered the natural flow of sediment and water around them.

Increased population has increased discharge of pollutants (nutrients, hydrocarbons, pathogens, and other agents) into shallow marine waters, adversely affect water quality, and hence the structure and function of marine communities. Man-made channels as well as natural embayments are subject to this additional stress. To reduce this stress, increased flushing is often desired, a result obtainable by dredging or by opening or maintaining new inlets. By increasing open bay water exchange and producing residual currents that are effective enhancers of tidal flushing, these activities can improve water quality. However, such multiple inlets may not be stable, desirable as they may be, and thus they may be expensive to maintain. This stability is the subject of the present volume: what happens when an embayment is served by multiple tidal inlets? Are multiple inlets an effective management tool? What scientific knowledge can be applied to answer these questions?

These and related concerns about water quality, storm flooding, navigation, and human health have led to pressure for increased management of coastal inlets. Residents surrounding bay waters serviced by coastal inlets often wish to maintain the quality and utility of those waters, even if expensive management options are required.

Since Bruun and Gerritsen's book, *Stability of Coastal Inlets*, (North-Holland, 123 pp.) was published in 1960, several landmark papers have examined the stability of tidal inlets. These articles (e.g., J. van de Kreeke, 1985, Stability of tidal inlets: Pass Cavallo, Texas. Estuarine, Coastal and Shelf Science, vol. 21, pp. 33-43) conclude that most multiple inlet systems are unstable: consideration of water exchange rates, critical inlet cross-sectional area, and sedimentation suggest that one or more of the inlets will eventually close. To derive the stability analysis, several approximations are made regarding the embayment geometry and flow characteristics. In this book, we examine some multiple inlet systems that do not fit the approximations of existing theory, and we present evidence of multiple inlets that can be stable on time scales of decades. New theoretical calculations support the evidence of stability in these cases.

The book was initiated following the opening of a new tidal inlet along Nauset Beach, in Chatham, Massachusetts, during a northeaster storm on 2 January 1987. This storm was not particularly damaging along much of the coast of New England, but had a profound influence on the community of Chatham. Since the time of the breach, property loss and threatened resources have caused strong divisiveness in the community, adversely affecting management of the threatened resources. Though scientists had warned a decade previously of the likelihood of such a breach, action was not taken to secure a consensus response when a breach did occur. Caught unaware or at least unprepared, Chatham has been struggling for the past five years to respond effectively to this threat to property and tax base, while balancing the positive effects such as increased tidal flushing, improved water quality, and improved navigation for fishing vessels. The future of the Chatham inlets, while hard to predict, is of concern to all. Science must help point the way to effective management.

The papers in this book discuss several aspects of multiple inlets. First, FitzGerald describes the origin and stability of tidal inlets in Massachusetts in general. The next series of five papers focus on the Chatham example: Liu et al. discuss the morphology and evolution of the new inlet; FitzGerald and Montello describe qualitative changes in the backbarrier region; Friedrichs et al. examine quantitatively the mechanisms leading to inlet formation and the stability of the multiple tidal inlets; Liu and Aubrey investigate altered transport pathways and residual currents from the three Chatham inlets; and finally, Weidman and Ebert examine the barrier spit which separates the north and south part of the Chatham embayment. Finally, a detailed analysis is presented of another stable multiple inlet system in Massachusetts, 20 km to the west of Chatham. In this paper, Aubrey et al. describe the complex planform of Waquoit Bay and the profound changes in circulation that have occurred as the system has alternatively been served by a single inlet, twin inlets, and finally, after Hurricane Bob in 1991, three inlets.

We wish to thank those who made this volume possible. We thank the reviewers for their helpful comments; the authors who have displayed patience while this volume was put together; and Pamela Barrows for typing and editing this entire volume. This work is a result of research sponsored by NOAA National Sea Grant College Program Office, Department of Commerce, under Grant No. NA86-AA-DSG090, Woods Hole Oceanographic Institution Sea Grant Project No. R/P-30-PD. We also would like to thank the Town of Chatham, the Commonwealth of Massachusetts' Coastal Zone Management Program, Friends of Pleasant Bay, the Woods Hole Oceanographic Institution's Sea Grant program, and the U.S. Army Corps of Engineers, all of whom have contributed to ongoing studies in this area.

David G. Aubrey
Graham S. Giese
Editors

1

Origin and Stability of Tidal Inlets in Massachusetts

Duncan M. FitzGerald

Abstract

The origin, morphology and sedimentation processes of tidal inlets along the Massachusetts coast are highly variable due to a wide range in physical settings. The factors which have governed their development and contributed to these different morphologies include wave and tidal energy, sediment supply, origin of the backbarrier, bedrock geology, sea level history, storms, and modifications by man. Some of the variability of these individual parameters can be related to the glacial history of this region. With the use of examples, physical data sets and morphological case histories, this paper examines the evolution and stability of inlets, primarily on the mainland coast of Massachusetts.

Introduction

The morphology and physical environment of the Massachusetts coast are as diverse as any comparable stretch of shoreline along North America. To a large extent the diversity has resulted from the varying effects of glaciation on a pre-existing, fluvially-eroded landscape. This has produced numerous

Formation and Evolution of Multiple Tidal Inlets
Coastal and Estuarine Studies, Volume 44, Pages 1-61
Copyright 1993 by the American Geophysical Union

large and small embayments, a wide range in shoreline orientations, and highly variable sediment supplies. This region encompasses the bedrock/till dominated shores of northwestern Buzzards Bay and Massachusetts Bay and the sandy coastal plain of southeastern Massachusetts and the region north of Cape Ann. Tidal inlets in this area are associated with many different coastal settings including large, well-developed barrier islands, long sandy spit systems, and narrow, transgressive sand and gravel barriers (Fig. 1). These inlets exhibit a wide range of sizes and morphologies which are related to their hydrographic regime, sediment abundance, bay size, tidal prism, and man-made modifications (Fig. 2). The origin of tidal inlets in Massachusetts is equally diverse encompassing riverine processes, barrier breaching, spit accretion, and other mechanisms.

During the past 10 years many harbors along the Massachusetts seaboard have filled to near capacity due to the dramatic increase in number of boat owners and their demand for boat slips. The existing overcrowded conditions coupled with future needs for harborage explains why the entrances to harbors are being maintained despite considerable cost to individual towns and State and Federal governments.More than half of the tidal inlets in Massachusetts have undergone modification projects to improve their navigation. In addition to navigation concerns, it is important to consider how the shoaling and closure of tidal inlets impact shellfishing, the exchange of nutrients between the nearshore and bays and marshes, and the well-being of juvenile species of many fin fish that use the inlets and bays as nursery grounds. Knowledge of inlet processes as determined from field investigations and historical analyses is vital in managing these resources and planning for future coastal development.

Tidal inlets are defined as openings in the shoreline through which water penetrates the land thereby providing a connection between the ocean and bays, lagoons or marsh and tidal creek systems. The main channel of a tidal inlet is maintained by tidal currents (Bruun and Gerritsen, 1955). The second half of this definition distinguishes tidal inlets from large, open embayments or passageways along rocky coasts. Tidal currents at inlets are responsible for the continual removal of sediment dumped into the main channel by wave action. Thus, according to this definition tidal inlets occur along sandy (or sand and gravel) barrier coastlines, although one side of an inlet may abut a

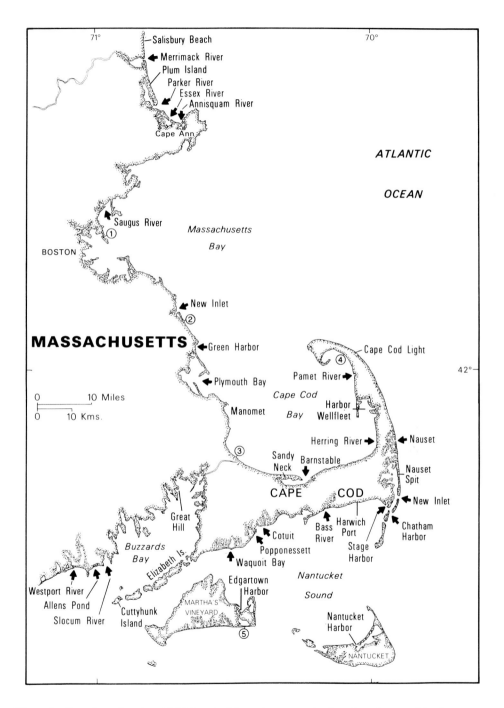

Figure 1. Location of major tidal inlets in Massachusetts and other sites discussed in the paper. Numbers 1-5 refer to locations where inlets have closed: 1. Shirley Gut, 2. South River Inlet, 3. Scusset Mills Creek Inlet, 4. East Harbor Inlet, and 5. Katama Bay Inlet.

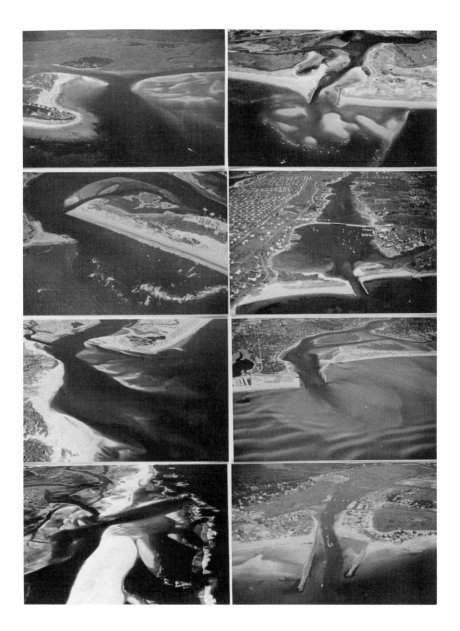

Figure 2. Oblique aerial photographs of: A. New Inlet, Scituate, B. Westport River Inlet, Westport, C. Parker River Inlet, Ipswich, D. Nauset Inlet, Eastham, E. Pamet River Inlet, Truro, F. Green Pond Inlet, Falmouth, G. Bass River Inlet, Dennis/Yarmouth, H. Green Harbor Inlet, Marshfield.

bedrock headland. Sand removed from the inlet channel is carried into the bay during the flood cycle forming flood-tidal deltas or transported seaward during the ebb phase forming ebb-tidal deltas (Fig. 3). The presence or absence of these sand shoals, their size, and how well they are developed are related to the region's tidal range, wave energy, sediment supply, and backbarrier setting. The general morphology of tidal inlets and their associated sand bodies and the processes that control sediment transport patterns are discussed in a review chapter by Boothroyd (1985) and in a recent volume on tidal inlets edited by Aubrey and Weishar (1988).

This paper will discuss the origin and variability of tidal inlets in Massachusetts and will demonstrate how natural and man-made changes to inlets affect their stability. Tidal inlet terminology will follow that of Hayes (1975, 1979).

Physical Environment

To understand the varying morphology, processes, and behavior of tidal inlets in Massachusetts, it is important to evaluate them in terms of the physical environment in which they have evolved. The morphological variability that exists along the Massachusetts coast can be explained in terms of an area's geological setting and hydrographic regime (Fig. 4). The glacial history of a particular shoreline segment dictates the sediment supply to the region and whether the coast is rocky or not. Wave energy and tidal range of the area influence how the sediment within the shoreline segment is dispersed. Major storms and the wind regime of the area also affect the pathways of sediment transport. Wave and tidal energy along the Massachusetts coast is largely controlled by the exposure of the shoreline and where it is situated with respect to major coastal bays.

Tides

The coast of Massachusetts can be divided into a number of shoreline segments and embayments based on similar tidal range (NOAA, 1991; Fig. 4). The region including Cape Cod and Massachusetts Bays and extending northward to the New Hampshire border is mesotidal (2.0 < TR < 4.0 m) with

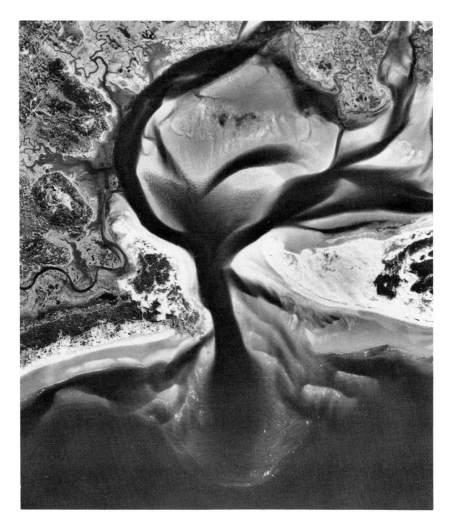

Figure 3. 1976 vertical aerial photograph of Essex River Inlet illustrating the morphology of ebb and flood-tidal deltas.

mean ranges between 2.5 and 3.1 m and spring ranges increasing to as much as 3.5 m at Wellfleet Harbor. Along the outer coast of Cape Cod the mean tidal range gradually diminishes to the south from 2.7 m at Cape Cod Light to 2.0 m at Chatham Harbor Inlet. This trend continues along Monomoy Island such that at Monomoy Point the mean tidal range is 1.1 m. Within Nantucket and Vineyard Sounds, including along the islands of Martha's Vineyard and Nantucket, the tides are microtidal (TR < 2.0 m) and generally the range decreases from east to west. At Harwich Port the mean range is 1.0 m and at Falmouth Heights 0.4 m. The shoreline in Buzzards Bay is also microtidal

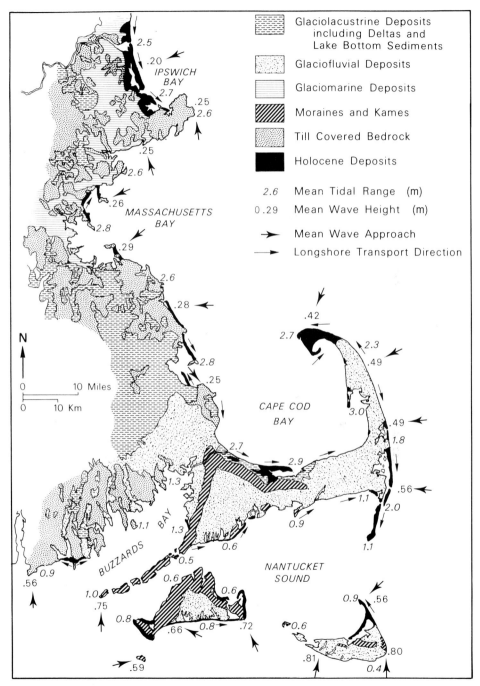

Figure 4. Physical setting of the Massachusetts coast including surficial deposits (from Larson, 1980; Stone and Peper, 1980), mean tidal range (from NOAA's tidal tables of North America), shallow water mean wave heights and dominant wave approach direction (from Jensen, 1983), and net longshore transport directions determined from spit growth, erosional-depositional trends in the vicinity of coastal structures and other coastal features.

with slightly larger ranges than the sounds to the east. At Great Hill near the entrance to the Cape Cod Canal the mean tidal range is 1.2 m and at Cuttyhunk Island near the entrance of Buzzards Bay the range is 1.0 m.

Given this distribution in tidal ranges, it can be expected that tidal inlets are larger, deeper and more stable along sandy shorelines where tidal ranges are large and bay areas are expansive. A comparison of the large, deep inlets north of Cape Ann versus the shallow inlets of Nantucket Sound illustrates this relationship well (Tables 1 and 2).

Waves

The highly variable orientation of the Massachusetts shoreline coupled with its numerous embayments causes different exposures to incident wave energy (Fig. 4). Temporal variations in wave energy are due to the seasonal distribution of storms and changing prevailing wind regime. Deepwater wave energies for this coast are known from a wave hindcast study for the region offshore of Nauset Beach, Cape Cod (U.S. Army Corps of Engr., 1957) and from a wave gauge located west of Cuttyhunk Island (Thompson, 1977). The shallow water wave energy (depth = 10.0 m) for the Massachusetts coast has been determined for 19 stations using 20 years of hindcast data (Jensen, 1983). The deepwater hindcast data indicate that the outer coast of Cape Cod and the shoreline to the north are dominated by east-northeast wave energy associated with the passage of extra-tropical northeast storms. The shallow water wave data corroborate this general trend with some exceptions due to sheltering and wave refraction processes.

The wave gauge off Cuttyhunk, which recorded three partial years of data, indicates that the deep water mean significant wave for this region is 0.9 m and the wave period is 7.5 sec (Thompson, 1977). The shallow water wave data for the southward facing shorelines show that the dominant wave energy comes from the south and that the south shores of Martha's Vineyard and Nantucket experience the largest waves along the Massachusetts coast (Jensen, 1983).

Wave energy within Cape Cod Bay, Buzzards Bay, Nantucket Sound, and Vineyard Sound is low due to limited fetch. Thus, wave processes along these coasts are tied closely to local wind conditions. The northern shores of Buzzards Bay and Nantucket and Vineyard Sounds experience greatest wave energy when extratropical storms or hurricanes pass to the west of Massachusetts generating strong southerly winds. Prevailing southerly winds also occur in these regions during the spring, summer, and early fall months. The southeastern coasts of Cape Cod Bay, Buzzards Bay, and Vineyard Sound are influenced by waves generated by prevailing northwest winds during the late fall, winter, and early spring months (Magee and FitzGerald, 1980).

The magnitude and direction of longshore sediment transport along the Massachusetts coast are highly variable and have been estimated from local erosional-depositional patterns around coastal structures, migration of inlets, growth of spits, and grain size trends. Net longshore transport directions are summarized in Figure 4.

Sediment Supply

The Pleistocene Epoch dictated the sediment distribution and abundance along the Massachusetts coast (Fig. 4) (Larson, 1982; Stone and Peper, 1982). Reworking of the glacial deposits produced the sand supply that was responsible for the development of the present day barrier and tidal inlet system. North of Cape Ann, the major source of sand for the coastal zone has been reworking of the Merrimack River delta that was deposited during the relative sea-level low stand, approximately 10,500 yrs BP (Edwards, 1988). These sediments have formed an extensive barrier system that extends from Great Boars Head, New Hampshire to Cape Ann.

The coastal region from Cape Ann south to Manomet is mostly sediment-starved containing exposures of bedrock with thin till covers (1 to 3 m thick) and some glacial marine deposits. Sediment is slightly more abundant in the vicinity of Boston Harbor and the South Shore where drumlins comprise much of the shoreline. The drumlins have a sand content of 30-40% (Newman et. al., 1990).

Table 1. Characteristics of selected tidal inlets along the Massachusetts coast.

Name of Inlet	Location	Setting	Associated Barriers	Present Backbarrier Environment	Associated Rivers and Streams	Inlet Mode of Formation
North Shore						
Merrimack River	Newburyport	Between Barrier Islands	Salisbury Beach, and Plum Island	Estuary and Marsh System	Merrimack River	Development of Regressive Barriers
Parker River	Ipswich	Between Island and Barrier Spit, and Controlled by Drumlins	Plum Island and Castle Neck	Marsh and Tidal Creeks with Open Water Areas	Parker and Ipswich Rivers	Development of Regressive Barriers
Essex River	Essex	Between Barrier Spit and Bedrock Outcrop	Castle Neck, and Coffin Beach Areas	Marsh and Tidal Creeks with Open Water Areas	Essex and Castle-neck Rivers	Development of Regressive Barriers
Annisquam River	Gloucester	Between Bedrock Outcrops	Wingaersheek Beach	Marsh and Tidal Creeks	Annisquam and Jones Rivers	Development of Regressive Barrier Beach
Saugus River	Revere/Lynn	Between Mainland and Barrier Spit	Point of Pines Spit	Marsh and Tidal Creek	Saugus and Pines Rivers	Spit Accretion
South Shore						
New Inlet	Scituate	Between Barrier Spit and Drumlin	Third Cliff Spit, and Hummarock Beach	Marsh and Tidal Creeks	North and South Rivers	Storm Breaching of Barrier, 1898
Green Harbor	Marshfield	Between Bedrock Outcrop and Barrier Spit	Green Harbor Spit	Marsh and Tidal Creeks	Green Harbor Rivers	Spit Accretion
Plymouth Bay	Plymouth	Between Drumlin and Barrier Spit	Duxbury Beach, Saquish Neck, and Plymouth Spit	Bay with Peripheral Marsh and Some Tidal Flats	Jones River	Spit Accretion
Cape Cod Bay						
Barnstable Harbor	Barnstable	Between Barrier Spit and Mainland	Sandy Neck	Open Water Areas and Marsh and Tidal Creeks	No Major Streams	Spit Accretion
Sesuit Harbor	East Dennis	Between Mainland and Small Barrier Spit	Sesuit Beach	Open Water Areas and Marsh and Tidal Creeks	Assessment Creek	Spit Accretion
Herring River	Eastham	Between Mainland and Barrier Spit	First Encounter Beach	Marsh and Tidal Creeks	No Major Streams	Spit Accretion
Pamet River	Truro	Between Two Barrier Spits	Harbor Bar Beach	Tidal Flats, Marsh, and Tidal Creeks	Pamet River	Spit Accretion
Outer Cape Cod						
Nauset Inlet	Eastham	Between Two Barrier Spits	Nauset Beach	Marsh and Tidal with some Open Water Areas	No Major Streams	Spit Accretion
New Inlet	Chatham	Between Barrier Spit and Barrier Island	Nauset Spit/ Nauset Island	Bay with Intertidal Flats	No Major Streams	Storm Breaching of Barrier, 1987
Chatham Harbor	Chatham	Between Two Barrier Islands	Nauset Island/ Monomoy Island	Bay with some Intertidal Flats	No Major Streams	Spit Accretion
Monomoy Breach	Chatham	Between Two Barrier Islands	Monomoy Island	Bay	No Streams	Storm Breaching of Barrier, 1978
Nantucket Sound						
Stage Harbor	Chatham	Between Two Barrier Spits	Harding Beach, Morris Island Dike and Spit	Bay with Tidal Flats	No Major Streams	Artificial Breach, 1945
Bass River	West Dennis/ Yarmouth	Between Mainland and Barrier Spit	West Dennis Beach Spit	Marsh and Tidal Creeks	Bass River	Spit Accretion
Cotuit Inlet	Barnstable	Between Mainland and Barrier Spit	Oyster Harbor Beach Spit	Bay	Mills River	Spit Accretion
Popponesset Bay	Barnstable/ Mashpee	Between Mainland and Barrier Spit	Popponesset Spit	Bay	Santuit and Mashpee Rivers	Spit Accretion
Waquoit Bay	Mashpee/ Falmouth	Between Two Barrier Spits	South Cape Beach	Bay	Quashnet River	Spit Accretion
Green Pond	Falmouth	Between Two Barrier Spits	Unnamed Spits	Bay	No Major Streams	Artificial Breach, 1951
Buzzards Bay						
Slocum River	South Dartmouth	Between Bedrock Outcrop and Barrier Spit	Slocum Spit	Bay with Peripheral Marsh	Slocum River	Spit Accretion
Allens Pond	South Dartmouth	Between Two Barrier Spits	Little Beach/ Allens Pond Spit	Bay with Marsh and Tidal Flats	No Major Streams	Artificial Breach, 1986
Westport River	Westport	Between Bedrock Outcrop and Barrier Island	Horseneck Beach	Estuary with some Marsh and Tidal Flats	East and West Branch of Westport River	Development of Regressive Barrier

Table 2. Morphology and stability of selected tidal inlets along the Massachusetts coast.

Name of Inlet	Location	Structure and Improvements	Stability	Inlet Dimensions Depth	Width	Flood Deltas	Ebb Deltas
North Shore							
Merrimack River	Newburyport	Double Jetties	Prior to Jetty Construction History of Southerly Migration and Breaching Back to North	10.1	323	Well-Developed, Intertidal	Subtidal
Parker River	Ipswich	None	Outer Channel Migrates South, Throat Stable	9.7	926	Well-Developed, Intertidal	Well-Developed, Intertidal
Essex River	Essex	None	Stable	12.2	354	Well-Developed, Intertidal	Well-Developed Sub/Intertidal
Annisquam River	Gloucester	Dredged Outer Channel	Stable	9.4	343	None	Subtidal/ Intertidal
Saugus River	Revere/Lynn	Revetments along Inner Channel	Stable	20.0	350	None	None, Modification by Man
South Shore							
New Inlet	Scituate	None	Stable in Present Location	8.5	230	None	Moderately Well, Subtidal
Green Harbor	Marshfield	Double Jetties and Dredged Channel	Channel Shoaling	8.0	140	None	None
Plymouth Bay	Plymouth	None	Stable	20.0	2000	Well-Developed, Intertidal	Well-Developed, Sub/Intertidal
Cape Cod Bay							
Barnstable Harbor	Barnstable	None	Stable	12.8	1400	Well-Developed, Intertidal	Well-Developed, Sits on Shallow Shelf
Sesuit Harbor	East Dennis	Double Jetties and Dredged Channel	Channel Shoaling	2.8	83	None	None
Herring River	Eastham	None	Stable	1.0	35	Well-Developed, Intertidal	Well-Developed, Sub/Intertidal
Pamet River	Truro	Double Jetties and Dredged Channel	Significant Channel Shoaling	2.3	90	Well-Developed, Intertidal	Well-Developed, Intertidal
Outer Cape Cod							
Nauset Inlet	Eastham	None	History of Northerly and Southerly Migration	3.2	265	Well-Developed, Intertidal	Well-Developed, Intertidal
New Inlet	Chatham	None	Still Equilibrating (see other papers this volume)	5.0	150	Well-Developed, Intertidal	Well-Developed, Mostly Subtidal
Chatham Harbor	Chatham	None	History of Southerly Migration	6.0	700	Moderately Well, Intertidal	Well-Developed, Sub/Intertidal
Monomoy Breach	Chatham	None	Continued Shoaling	2.0	220	Well-Developed, Intertidal	Poorly Developed, Subtidal
Nantucket Sound							
Stage Harbor	Chatham	Dredged Outer Channel	Channel Shoaling	3.0	80	None	Well-Developed, Intertidal
Bass River	West Dennis/ Yarmouth	Double Jetties and Dredged Channel	Channel Shoaling	3.0	130	None	Subtidal, Sits on Platform
Cotuit Inlet	Barnstable	Dredged Outer Channel	Channel Shoaling	3.4	240	Well-Developed, Subtidal	Moderately Well, Subtidal
Popponesset Bay	Barnstable/ Mashpee	None	History of Northerly Migration and Breaching Back to North	2.0	75	Well-Developed, Intertidal	Poorly Developed, Intertidal
Waquoit Bay	Mashpee/ Falmouth	Double Jetties	Channel Shoaling	3.0	35	Moderately Well, Sub/Intertidal	Poorly Developed
Green Pond	Falmouth	Double Jetties and Dredged Channel	Channel Shoaling	2.1	80	Intertidal	Subtidal
Buzzards Bay							
Slocum River	South Dartmouth	None	Widening Due to Spit Erosion	3.4	90	Moderately Well, Sub/Intertidal	Intertidal, Sits on Intertidal Platform
Allens Pond	South Dartmouth	Occasional Artificial Breaching of Spit when Inlet Closes	Migrating Westward	1.0	60	None	Poorly Developed, Subtidal
Westport River	Westport	Dredged Channel	Stable	7.6	260	Intertidal	Subtidal

South of Manomet, including most of Cape Cod and much of the Martha's Vineyard and Nantucket shorelines, sand is abundant due to the presence of extensive glacial outwash deposits. Areas with less sand resources coincide with coasts composed of glaciolacustrine deposits (e.g., parts of southern Cape Cod Bay). or moraine deposits (e.g., northern shore of Martha's Vineyard and the Elizabeth Islands).

The northern shore of Buzzards Bay is also sediment starved and is character-ized by till covered peninsulas separated by deep embayments (FitzGerald et. al., 1987). Sediment is slightly more abundant along the southwestern half of the shoreline due to the presence of some glaciofluvial and glaciolacustrine deposits in addition to some thicker till deposits such as Gooseberry Neck, a drumloidal feature offshore of Horseneck Beach.

Occurrence of Tidal Inlets

Introduction

The formation of a tidal inlet requires the presence of an embayment and the development of barriers. In coastal plain settings, often the embayment or backbarrier is formed through the construction of the barriers themselves, like much of the East Coast of the United States or East Friesian Islands of the North Sea. In Massachusetts, the origin of the embayment may be related to drowned river valleys, rocky or sandy irregular coastlines, kettles, groundwa-ter sapping channels, or the formation of a barrier chain. In these settings, tidal inlets are formed when the opening to the embayment becomes constricted by barrier construction across the embayment or when an existing barrier is breached during a storm or cut artificially. Various settings of tidal inlet development in Massachusetts are listed in Table 1 and discussed below.

Drowned River Valleys

The best example of tidal inlet development in a drowned river valley setting is Merrimack River Inlet located between Salisbury Beach and Plum Island

(Fig. 1). The Merrimack River, which drains much of New Hampshire and northeastern Massachusetts, delivered a large quantity of sand to the coastal region during deglaciation. Much of this sediment was deposited in the form of three major deltas at 33 m and 16 m above present mean sea level and 50 m below mean sea level (Edwards, 1988). The last delta was deposited during the Holocene lowstand and was formed, in part, through the cannabalism of the 16 m elevation delta (Edwards, 1988). Subsequent drowning of this erosional valley during the late Holocene formed the present day embayment at the river mouth. Later, the embayment was constricted during the evolution of Plum Island and Salisbury Beach, resulting in the formation of Merrimack River Inlet. The major sand source for these regressive barriers and the barriers to the south was the onshore reworking of the top portion of the 50 m delta during the Holocene transgression.

The formation of Plymouth Bay and location of its entrance are also closely related to deglaciation processes (Fig. 5). As the Buzzards Bay Lobe of the continental ice sheet retreated northward across southeastern Massachusetts, sometime after 15,300 yrs BP (Larson, 1982), glacial Lake Taunton was formed covering an area of approximately 140 km². During much of its existence the lake drained to the south through a spillway just north of Fall River (Larson, 1982). However, after the Cape Cod Bay Lobe retreated northeastward removing the lake's eastern dam, the water drained through the Jones River valley, which was 4 to 7 m lower than the lake level (Larson, 1982). Presently, the river forms the estuarine headwaters of Kingston Bay within Plymouth Bay (Fig. 6). The greatest thickness of sediments above the acoustic basement (> 20 m) in the Plymouth Bay area is along two troughs; one coinciding with the north-south long dimension of the bay and the other defining the present course of Plymouth Inlet's main ebb channel (Hill et al., 1990; Fig. 6). This inferred paleodrainage system inside the bay joins with the Postglacial drainage patterns outside the bay as reported by Oldale and O'Hara (1977). Thus, it would appear that drainage established during the early Holocene has dictated, to a large extent, the geometry of Plymouth Bay and the position of its inlet. The barriers that front Plymouth Bay have evolved from landward migrating transgressive barriers and through spit accretion from sediment eroded from nearby drumlins and till cliffs (Hill and FitzGerald, in press).

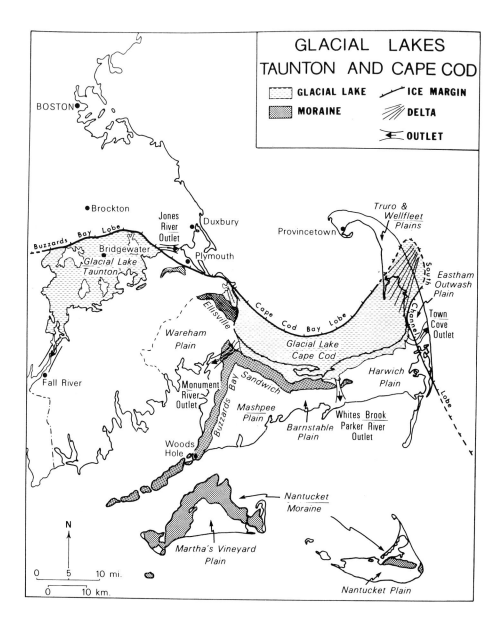

Figure 5. Map of southeastern Massachusetts depicting deposits and features of late Wisconsinan glaciation (modified from Larson, 1982).

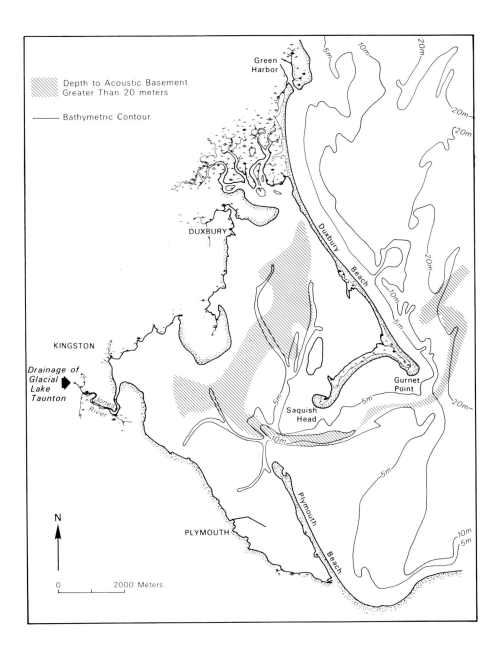

Figure 6. Map of Plymouth Inlet and embayment showing bathymetry and structure contours of the depth to the acoustic basement. Note that the northern portion of the bay and the inlet channel coincide with where sediment thickest is greatest. The two arms are believed to be major channels that were active when Glacial Lake Taunton was draining to the east (after Hill et al., 1990).

Rocky Irregular Shorelines

The Cape Ann promontory and the northern coast of Buzzards Bay are the major rocky coastlines in Massachusetts. Along these shorelines pocket beaches are the dominant accretionary landform and tidal inlets exist only where the sediment supply was abundant enough to develop significant barriers (Fig. 3). The lack of sand along Cape Ann has prohibited barrier and inlet development except for a small inlet associated with the pocket barrier of Good Harbor Beach. Sediment is slightly more abundant along the northwest coast of Buzzards Bay and tidal inlets are more numerous (Fig. 3; Table 1). This shoreline is characterized by deeply incised embayments fronted mostly by thin, sand and gravel barriers. In this region inlets were formed by spit accretion derived from sediment that had been eroded from adjacent shorelines, as well as sediment moved onshore from nearshore glacial deposits (FitzGerald et al., 1987). The larger inlets in this area, including Slocum and Westport River Inlets, are positioned next to bedrock outcrops. Several inlets along this coast have closed in historical times due to the transgression of the barriers.

Sandy Irregular Shorelines

The original Cape Cod shoreline that was formed by rising sea level during late Holocene (approximately 3,000 to 4,000 yrs BP) was probably highly irregular due to the nonuniform topography of the moraines, outwash plains and other glacial sediments that comprise Cape Cod. It is likely that proto-Cape Cod had the same general "arm" form but with numerous embayments and small islands (Davis, 1896). This shoreline has been smoothed through erosion of headlands, disappearance of some of the islands, and development of spits across the bays. Numerous tidal inlets were formed as a result of spit accretion including Barnstable Harbor Inlet, Nauset Inlet, Chatham Harbor Inlet, and many others (Fig. 1; Table 1).

A basal peat sample collected at a depth of 5 m below the present marsh surface at Scorton Neck near the beginning of Sandy Neck was radiocarbon dated at 3,170 yrs BP (Fig. 7; Redfield, 1967). This date and others were used by Redfield (1967) to hypothesize that Sandy Neck began forming not more than 4,000 yrs BP. The sand that comprises Sandy Neck was eroded from the Wareham Pitted Plain, Ellisville Moraine and other surrounding glacial deposits (Fig. 5) and transported south by littoral processes. The sandy cliffs along the Manomet and Sagamore shoreline are evidence of this erosion. It is likely that the formation of other tidal inlets along Cape Cod also occurred approximately 3,000 to 4,000 yrs BP, coincident with rising sea level and spit accretion. However, the barriers that front the other inlets are considerably younger than Sandy Neck due to the transgressive nature of most of them. This has been documented at various inlets on Cape Cod by FitzGerald and Levin (1981), Aubrey and Gaines (1982), Aubrey and Speer (1984), and Giese (1988).

Kettles

A unique means of coastal bay development and tidal inlet formation occurred along the Cape Cod Bay shoreline in Eastham. This portion of Cape Cod is composed of the Eastham Outwash Plain (Fig. 5) which contains numerous kettles. One of the largest of these kettles (1,200 m across) is located on the coast and forms the embayment behind Herring River Inlet (Fig. 8). Thirty-six auger cores taken throughout the marsh system landward of First Encounter Beach indicate that the base of the kettle is at least 8 m deep (Fig. 9). The cores reveal that the marsh peats and organic muds are thickest in the eastern side of the embayment and thin toward the inlet mouth and barrier spit. The marsh deposits are underlain by medium-to-coarse sands that are moderately well-sorted. The western third of the embayment contains little or no marsh deposits at the surface and is covered by supratidal vegetation (Fig. 9). The shallowness of the cores in this region does not allow for a determination of the presence of marsh peats at depth (> 2.5 m).

The stratigraphy of the kettle and morphology of the present barrier spit and tidal inlet system suggest that during the Holocene transgression, rising sea level flooded the kettle forming a large embayment (Fig. 10). Sand eroded from the coast to the north and transported south built a spit across the mouth of the embayment forming Herring River Inlet. A scenario for the filling of the bay begins with the contemporaneous deposition of sediment along the fringe of the bay with marsh growth toward the center, and the deposition of

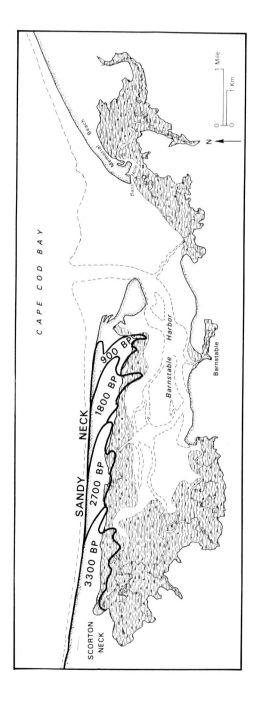

Figure 7. Chronology of the growth and development of Sandy Neck. The sediments that built this spit were derived from the erosion of glacial cliffs to the north (after Redfield, 1967).

Figure 8. Oblique aerial photograph of Herring River Inlet and marsh system. This inlet is located on the pitted Eastham Outwash Plain. The bay of this inlet was originally a kettle that became connected to the sea.

sheet sands and flood-tidal deltas along the seaward side of the embayment. Storm waves overwashing First Encounter Beach during events like the Blizzard of 1978 would have introduced large quantities of sand into the bay and may explain the lack of surface peat and organic mud deposits in the eastern third of the embayment. Sediment deposited along the margin of the bay would have come from overland sources and from fine-grained sediments carried in suspension by tidal currents. As the bay was converted to high marsh with small tidal creeks, the tidal prism was greatly reduced, resulting in a smaller equilibrium inlet cross section and elongation of the spit system. Spartina marsh peats cropping out in the intertidal zone seaward of First Encounter Beach suggest that the decrease in bay area is also a result of the transgression of First Encounter Beach.

Groundwater Sapping Channels

One of the noteworthy coastal morphologies along the south shores of Cape
Cod, Martha's Vineyard, and Nantucket is the north-southward trending,
flooded valleys that form the inlet-associated bays of this region (Fig. 11).
The depressions that resulted in the formation of these elongated bays were
once considered to have originated from meltwater streams (FitzGerald,
1985) due to permafrost conditions (Oldale and Barlow, 1986); recent work

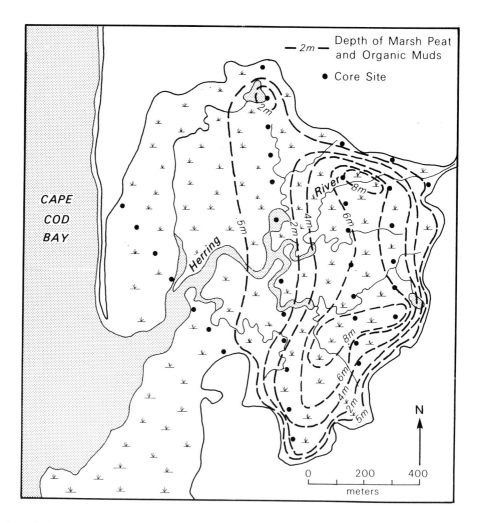

Figure 9. Location of auger cores and thickness of marsh peats and bay-fill mud deposits at Herring
River backbarrier region.

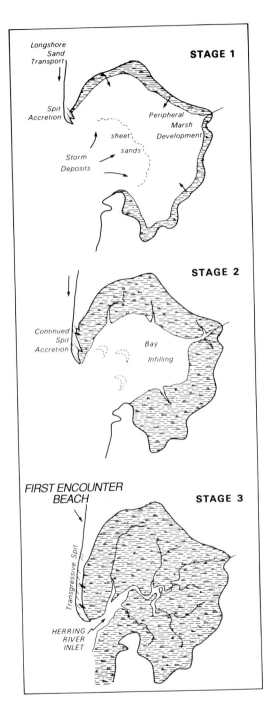

Figure 10. Conceptual model of tidal inlet formation and marsh development of Herring River

suggests that they developed through the process of groundwater sapping (Caldwell, pers. comm.; D'Amore, 1983). Topographic maps show that the bays are fairly evenly spaced along a given stretch of shoreline and have a pinnate drainage structure which is unlike the pattern that would have resulted if the valleys formed from a braided stream network associated with an outwash plain. Secondly, it is reasonable to assume that the hydraulic head produced by Glacial Lake Cape Cod (Fig. 5), which was at least 29 m (Larson, 1982; Oldale, pers. comm.), coupled with the coarse-grained Mashpee out-wash plain would have caused piping as groundwater flowed toward the depression which is now Nantucket Sound. This process is known to move sand-sized material and create channels that migrate headward as they develop (D'Amore, 1983). The draining of Lake Cape Cod would have terminated this process and rising sea level would have eventually flooded the stream valleys. During the same period, sand that eroded from the intervalley headlands would have fed spit systems that built across the flooded bays forming tidal inlets.

Barrier Chains

The Massachusetts coast has two major barrier chains; one extending from Great Boars Head in New Hampshire to Cape Ann and another that stretches along the outer coast of Cape Cod from Coast Guard Beach to Monomoy Island (Fig. 12). The mode of inlet formation along these two chains was quite different and related to differences in barrier development and river drainage patterns.

Outer Cape Cod Chain

The barriers forming the Nauset Spit-Monomoy Island chain formed through spit accretion from sediment eroded mostly from the glacial cliffs north of Coast Guard Beach (Fig. 12) (Fisher, 1987; Giese, 1988 and this volume). Periodically, storm breaching has segmented these barrier spits, such that at various times there are two or more quasi-stable inlets. Quite recently Monomoy Island was breached during the 6-7 February Blizzard of 1978 and Nauset Beach was breached during the northeast storm of 2 January 1987.

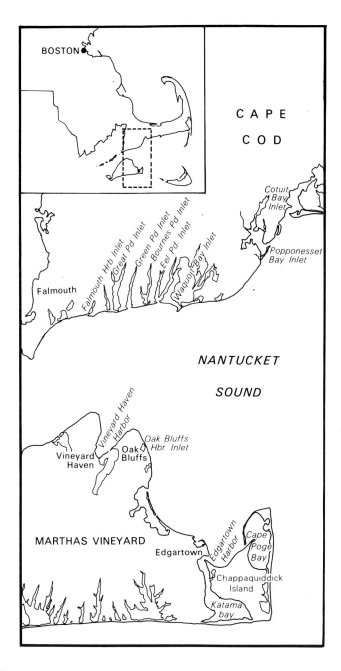

Figure 11. Groundwater sapping channels along the southern shores of Cape Cod and Martha's Vineyard. Inlets are unstable along the south shore of Martha's Vineyard due to small bay areas, small tidal ranges and moderate wave energy. On the southwest coast of Cape Cod similar conditions have necessitated the construction of jetties to keep inlets open and navigable.

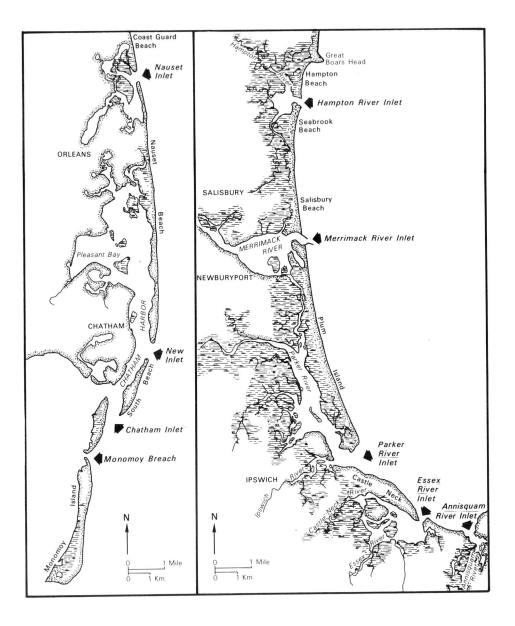

Figure 12. Major barrier coasts in Massachusetts.

The segmentation of the barriers and the development of inlets along this coast are related to a gradual restriction of tidal flow through existing inlets due to spit accretion and inlet migration (Giese, 1988; this volume). This produces differences in tidal range and tidal phase between the ocean and bays which can produce a substantial hydraulic head across the barrier. Under these conditions the barrier is susceptible to breaching, particularly during storms when the hydraulic head increases due to the storm surge.

Thinning of barriers is also a key factor in controlling when spits are breached. If the barrier is wide and has a well-developed frontal dune ridge and secondary dune system, breaching is difficult regardless of the hydraulic head. In contrast, destruction of the foredune ridge and thinning of the barrier allows barrier overwashing, channelization of the return flow, and subsequent inlet formation. Historical shoreline change data of the glacial cliffs north of Nauset Spit indicate that for the period between 1938 and 1974 there were significant temporal and spatial variations in shoreline location (Gatto, 1978). Thus, it can be reasoned that during the same period of time the supply of sand to the southern barrier system may have been equally variable, which may have influenced the retreat and advance of the barrier shoreline. Changes in the trend of shoreline retreat and advance are probably related to natural variations in wave energy and the frequency of major storms. Thus, breaching of the outer Cape Cod barrier system occurs when a sufficient hydraulic head has been established and the barrier has sufficiently thinned to facilitate overwashing during a major storm (see Friedrichs et al., this volume).

Northern Massachusetts Chain

The barrier chain north of Cape Ann contains five major barriers and five tidal inlets (Fig. 12). Although various workers have proposed littoral currents and spit accretion as responsible for the formation of these barriers (Nichols, 1941; McIntire and Morgan, 1964; Rhodes, 1973), these authors were unaware of the large accumulation of sand that exists in the Merrimack River delta (vol. = 1.4 x 10 m) 6 km offshore of the present river mouth in 50 m of water (Edwards, 1988). It is now believed that sand which formed this barrier chain came primarily from a reworking of the 50 m depth Merrimack marine delta and to lesser extent from the reworking of other glacial deposits on the continental shelf and some sediment discharged from the Merrimack River. Using the shallow seismic reconstruction of the 50 m delta by Edwards (1988), Som (1990) calculated that erosion and onshore transport of the top 2.5 m of the delta during the Holocene transgression could account for the entire volume of sand comprising the barrier chain and tidal delta shoals. It has been widely reported that marine deltas can be a significant source of sediment in development of barriers, including the coasts of Maine (Belknap, 1987; FitzGerald et al., 1990), North Carolina (Hine et al., 1979), Georgia (Oertel, 1979), and Louisiana (Penland et al., 1988).

It is believed that the present barrier chain began forming during the Mid-Holocene from transgressive barriers containing numerous ephemeral tidal inlets. In a stratigraphic study of northern New England, McIntire and Morgan (1963) dated the initial stage of Plum Island development as occurring sometime prior to 6,300 yrs BP. The other barriers to the north and south probably formed shortly thereafter from sand delivered onshore from the shelf and from sand moved alongshore by wave action. As the barriers stabilized and increased in width, tidal inlets probably decreased in number and also became more stable. Tidal inlets along this chain are associated with one or more river systems, although with the exception of the Merrimack River, they are small and discharge little freshwater compared to their saltwater tidal prisms (Table 1). Although the rivers are small, their valleys provided ideal locations for inlets to stabilize and the development of backbarrier marshes and tidal creeks. The association of tidal inlets with former river valleys is common along many barrier coastlines (Morton and Donaldson, 1973; Oertel, 1975; Halsey, 1979). Inlets along this chain are also partially stabilized or anchored next to bedrock outcrops (Hampton, Essex, Annisquam River Inlets).

Thus, the inlets along these two chains differ in that the Cape Cod inlets are associated with spit systems, are formed as a result of barrier breaching, and tend to migrate. In contrast, inlets north of Cape Ann are associated with a barrier coast that evolved from transgressive barriers, formed in paleo-river valleys, and are relatively stable.

Morphological Variability

Variability in tidal inlet morphology along the Massachusetts coast is a product of the vastly different physical settings under which inlets have formed and evolved. Tidal inlets may differ from one another in size and channel geometry, shoreline configuration, associated sand shoals, back-barrier setting and other components. Many of the major differences among the inlets can be explained in terms of varying wave and tidal conditions (Hayes, 1975; 1979). Sediment supply and tidal prism are other important variables that govern inlet morphology (Davis and Hayes, 1984). Character-istics of the tidal inlets discussed in this section are listed in Tables 1 and 2.

Inlet Size

The cross-sectional area of an inlet is dictated by its tidal prism (O'Brien, 1931; 1969) which, in turn, is primarily a function of bay size (open water area) and bay tidal range. The largest inlets in Massachusetts occur along mesotidal shorelines where backbarrier areas are expansive and composed chiefly of open water. Plymouth Inlet (Fig. 6) is such an inlet, having three large contiguous bays composed of open water areas and tidal flats. It has a spring tidal range of 3.3 m. These conditions combine to produce a spring tidal prism of 1.2×10^6 m^3 and an inlet cross-sectional area of 9,160 m^2 (Hill et al., 1990). Other large tidal inlets occur along the barrier chain north of Cape Ann (Merrimack, Parker, Essex Inlets) and in Cape Cod Bay (Barn-stable Harbor Inlet). These inlets have mesotidal ranges (TR = 3.0 m) and large backbarrier areas (Table 1).

Tidal inlets are relatively small along the microtidal shorelines of Buzzards Bay (TR = 1.0 to 1.3 m) and Nantucket (TR = 0.4 to 1.2 m) and Vineyard Sounds (TR = 0. 5 to 0.8 m). In these regions the low tidal ranges added to the diminutive size of most of the inlet associated bays result in small tidal prisms and small equilibrium inlet channels (Table 2). Even at Westport River Inlet (Fig. 2) which drains both the East and West Branch of the Westport River Estuary, the inlet throat is only 250 m wide with a stable cross-sectional area of 850 m and an average depth of 3.4 m (Magee and FitzGerald, 1980). In comparison, the bay areas of both the Parker River and Essex River Inlets, north of Cape Ann, are smaller than that of Westport River Inlet, however their inlet throat cross sections are more than twice as large (3,097 m² and 1714 m², respectively; FitzGerald, unpub. data).

Along the southern coast of Cape Cod the effect of small tidal ranges on inlet size is particularly well illustrated. Despite the relatively protected environment within Nantucket Sound, which produces low wave energy and small longshore transport rates (4,400 m³ net easterly transport in the vicinity of Bass River Inlet; Slechta and FitzGerald, 1982), most tidal inlets are jettied and/or dredged. Tidal prisms and tidal currents are insufficient along most of this microtidal shoreline to maintain navigable entrance channels for pleasure craft.

Associated Sand Shoals and Backbarrier Settings

Sand which is dumped into the inlet channel by littoral processes and flood tidal currents is transported seaward by ebb currents to the ebb-tidal delta or moved landward into bays forming flood-tidal deltas. Ebb-tidal deltas are links and short-term repositories in the littoral transport system that allow sand to bypass inlets. Flood-tidal deltas my build vertically to form intertidal sand shoals which subsequently may be colonized by marsh vegetation, resulting in the filling of the bay (Lucke, 1934). Models depicting tidal deltas and inlet settings were first put forth by Hayes et al. (1973) and Hayes (1975), originally based on the tidal range of the region. Later, these geomorphic models were modified to include the influence of wave energy (Hayes, 1979; Nummedal and Fischer, 1978).

Ebb-tidal deltas

In Massachusetts ebb-tidal deltas are well developed along mesotidal shorelines at medium-to-large inlets (Table 2). At these locations, like Essex River Inlet (Fig. 3), the ebb delta has a main ebb channel that incises a broad arcuate accumulation of sand called the swash platform (Hayes, 1975). On top of the swash platform are wave built swash bars which migrate onshore eventually attaching to the beach (Hine, 1975; FitzGerald, 1976). The main channel shoals in a seaward direction and is often bordered by linear bars. In mesotidal settings where sand is abundant, swash bars and channel margin linear bars are often exposed at low tide. At large jettied inlets like Merrimack River Inlet the ebb-tidal delta forms too far offshore for intertidal bars to develop. Likewise, at New Inlet along the South Shore (Fig. 2) the paucity of sand in this region probably prevents bars from building vertically to an intertidal exposure. In contrast, at the structured Pamet River Inlet where an abundant sand supply leaks around the updrift jetty, intertidal bars are well formed (Fig. 2). The small tidal prism and relatively weak tidal currents of this inlet result in the ebb delta forming in shallow water close to the inlet mouth (FitzGerald and Levin, 1981).

Ebb-tidal deltas are much more poorly developed along the microtidal shorelines of Buzzards Bay and Nantucket Sound (Table 2). In these regions the ebb delta is completely subtidal due to relatively small tidal prisms and smaller tidal range to expose the sand shoals. At many inlets, like Westport River Inlet (Fig. 2), the ebb delta is best defined during large wave conditions which serve to outline its extent. At other inlets, such as Slocum River Inlet and several inlets along the southern coast of Cape Cod (Bass River Inlet, Fig. 2), the ebb-tidal delta is moderately well developed and visible in aerial photographs because it has formed on a shallow nearshore platform. Ebb deltas at small tidal inlets along microtidal shores are mostly absent (Table 2).

Flood-tidal deltas

Most tidal inlets in Massachusetts have singular or multiple flood-tidal deltas, provided there is enough space in the backbarrier for them to form (Table 2). Flood deltas develop landward of the inlet throat where tidal current velocities diminish due to an increase in channel dimensions. At inlets where filling of the backbarrier has produced marsh islands and tidal creeks with little open-water area, flood deltas may be absent (e.g., New Inlet, Scituate; Fig. 2). In some instances, deltas become colonized and modified by marsh growth and are no longer discernible as flood-tidal delta landforms (cf., FitzGerald et al., 1990). At jettied inlets and inlets with boat marinas, flood deltas are often removed to provide better navigation or space for boat moorings (e.g., Green Harbor, Scituate, Fig. 2).

Flood-tidal deltas are normally horseshoe-shaped and consist of a flood ramp that bifurcates into flood channels through which sand is transported onto the delta platform (Fig. 3). The ebb shield which defines the landward extent of the delta is the highest part of the delta and is commonly partially vegetated by Spartina grasses. This part of the delta shields the rest of the shoal from effects of the ebb currents. Sand eroded from the ebb shield by ebb tidal currents is carried seaward forming ebb spits which extend toward the inlet throat (Boothroyd and Hubbard, 1975; Hayes, 1975).

On the Massachusetts coast, flood deltas are best developed at large inlets along mesotidal shorelines (Table 2). For instance, flood deltas are well formed with intertidal exposures at Merrimack, Parker, and Essex River Inlets north of Cape Ann and at Plymouth Inlet and Barnstable Harbor Inlet in Cape Cod Bay. There are multiple flood deltas at Nauset Inlet (Fig. 2); their presence influences the flow of water through the inlet and the pattern of inlet migration (Aubrey and Speer, 1984). Multiple flood deltas are still evolving landward of the breach through Nauset Beach and their resulting configuration and location will strongly affect the patterns of flow in Chatham Harbor and Pleasant Bay (FitzGerald and Montello, 1990; see other articles in this volume). A large flood delta on the western side of the Monomoy Breach is presently undergoing modification due to the recent closure of this inlet.

Along the microtidal shorelines of Buzzards Bay and Nantucket Sound, flood-tidal deltas are usually small compared to those found along mesotidal shorelines. Commonly, much of the delta is subtidal and irregularly shaped (e.g., Westport River Inlet and Green Pond Inlet; Fig. 2). Their diminutive nature probably is related to smaller tidal prisms and weaker tidal currents. Storms are a major cause of flood delta development along microtidal coasts resulting from the process of barrier breaching (Pierce, 1976) or increased sediment being delivered to the inlet coupled with elevated flood current strength associated with storm surge development (FitzGerald, 1988).

Backbarrier Systems

There are two major types of backbarrier environments associated with tidal inlets in Massachusetts and these correlate well with tidal range (Hayes 1975,

1979). Tidal inlets along mesotidal coasts have backbarrier areas composed primarily of high tide marsh (Spartina patens) incised by major and minor tidal creeks. At inlets north of Cape Ann and New Inlet in Scituate, rivers form the major tidal channel(s) in the backbarrier (Fig. 12 and 2, respectively; Table 1). In mesotidal settings the percentage of open water area and intertidal flats decreases away from the inlet mouth while the percentage of marsh increases (Fig. 13; Som, 1990).

In microtidal settings tidal inlets connect the ocean to shallow bays or lagoons (e.g., Green Pond and Bass River Inlet; Fig. 2, Table 1). In the case of Westport River Inlet, the bays are drowned river valleys with some intertidal flats and marsh areas. Most of the marsh islands occur behind the middle of Horseneck Beach at the site of an old tidal inlet and probably represent flood-tidal delta shoals that were deposited before the inlet closed (Magee and

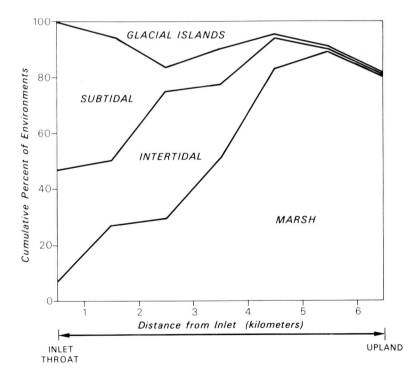

Figure 13. Distribution of backbarrier environments associated with Essex River Inlet. Note that with increasing distance away from the inlet mouth, the percentage of marsh increases while open water areas and intertidal flats decrease (after Som, 1990).

FitzGerald, 1980; Ibrahim, 1986; FitzGerald et al., 1987). The difference between the marsh and tidal creek backbarrier setting of mesotidal inlets versus the open water and fringing marsh of microtidal inlets is probably related to the larger tidal prisms, stronger tidal currents, and greater potential of bringing sediment into the bay at inlets with larger tidal ranges. The greater tidal fluctuation in the bay also produces larger intertidal areas which promote marsh formation and stabilization of fine-grained sediment. One exception to this trend is the mesotidal backbarrier system of Plymouth Inlet (Fig. 1) where the bay is composed principally of intertidal sand and mud flats and open water areas (Fig. 6). A detailed stratigraphic and sediment transport study of this region has revealed that Plymouth Bay has been a sediment sink since its formation during the Mid-to-Late Holocene and that the bay fill consists generally of a fining upward sequence of sands and muds (Hill et al., 1990; Hill and FitzGerald, in press). The presence of intertidal flats and absence of marshes indicate that these environments are not suitable for marsh development. This condition is likely the result of the size of the bay and tidal range which allow tidal and wave-generated currents at high tide and especially during storms to inhibit colonization of the flats by halophytic grasses. Ice gouging of flats during the winter may also be an operative process. The absence of salt marshes despite the presence of expansive tidal flats has been noted along the Friesian Islands in the North Sea and behind the Copper River Delta barrier system in Alaska (FitzGerald and Penland, 1987).

Tidal Inlet Stability

Stable tidal inlets are in dynamic equilibriun with the scouring action of tidal currents and the infilling of sediment delivered by longshore currents (Inman and Frautschy, 1965). However, the equilibrium of the inlet channel does not imply stability in position of the inlet, rather only in its cross-sectional area. The size of an inlet has been shown to be proportional to the volume of water flowing through it during a half tidal cycle (tidal prism). This relationship was quantified by O'Brien (1931, 1965) and later refined by Jarrett (1976) for structured versus unstructured inlets and inlets with varying wave energy (i.e., Pacific, Gulf and Atlantic Coasts).

The stability of inlets along the shores of Denmark, Netherlands and United States was examined in detail by Bruun and Gerritsen (1959) and Bruun (1967) and found to be governed by shear stress along the channel bottom. They noted that the magnitude of shear stress and maximum current velocities in the channel necessary to flush the inlet of sediment varied according to inlet geometry, rate of littoral drift delivered to the inlet, and concentration of suspended versus bedload. Later investigators suggested that inlets possess a critical cross-sectional area and if inlet size is reduced below this critical value through the influx of sand, it will close (O'Brien and Dean, 1972). While these various relationships would be useful in interpreting the evolution and closure of certain inlets in Massachusetts, the lack of hydraulic and morphologic data concerning these inlets make these analyses impossible. Therefore, the stability of Massachusetts inlets will be evaluated using historical information and other data sources. Effects of varying tidal prism and wave energy, changes in sediment supply, inlet closures and openings, and jettied inlets will be examined.

Tidal Prim and Wave Energy

The influence of tidal prism and wave energy on the equilibrium cross sectional area of tidal inlets is illustrated well along the northwest coast of Buzzards Bay. This shoreline consists of elongated bays fronted by thin transgressive barriers; the beach ridge barrier of Horseneck Beach at Westport River Inlet is a major exception. As shown in Figure 14, there is a close correspondence between bay size and inlet width, with large bays having wider inlets. This relationship exists because tidal range is fairly constant along this coast and bay area can be taken as a first order approximation of inlet cross-sectional area. The fact that many of the bays have no permanent connection to the sea is a function of a limited sediment supply in a regime of rising sea level (Fig. 4). During the ongoing transgression the lack of sediment along most of this coast has resulted in a landward migration of the barriers, a process which is decreasing bay area at a faster rate than the upland has been inundated by rising sea level. This has reduced the tidal prisms of many of the bays causing the closure of the smaller inlets. This same phenomenon explains the lack of tidal inlets along the elongated pond shorelines of Martha's Vineyard and Nantucket (Fig. 11). If jetties had not been constructed at many of the inlets along the Cape's Nantucket Sound, several of them would have closed.

The relationship depicted in Figure 14 also illustrates the importance of wave energy in influencing the stability of inlets. Note that while tidal inlets exist at Salters Pond and Little River Inlet, the larger bays of Quicksand Pond and Briggs Marsh Pond, which potentially would produce larger tidal prisms than the other two, maintain no permanent inlets. This apparent contradiction in the aforementioned area/inlet width relationship can be explained due to differences in wave energy. The eastern two bays that have tidal inlets (Salters Pond and Little River Inlet) are partially sheltered from wave energy by headlands and the offshore Elizabeth Islands. In contrast, the barriers fronting Quicksand and Briggs Marsh Ponds are directly exposed to the prevailing southerly wave climate (Fig. 4). Thus, for inlets along the Buzzards Bay coast that are close to the condition which produces instability and closure (O'Brien and Dean, 1972), it appears that slight differences in

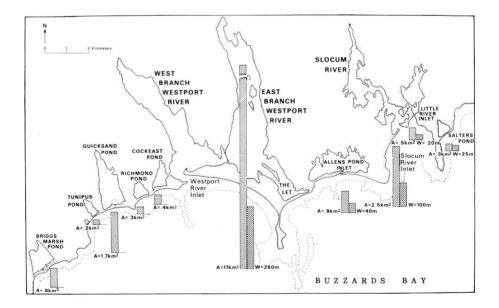

Figure 14. Plot of bay area versus tidal inlet width for the northwest Buzzards Bay coast. This diagram illustrates the relationship between tidal prism and inlet cross sectional area. Larger bay areas produce larger tidal prisms which require larger tidal inlet openings and conversely (from FitzGerald, 1988).

wave energy and longshore sediment transport rates can control the fate of the inlet (FitzGerald et al., 1987).

Changes In Sediment Supply

Along barrier island coasts a decrease in sediment supply leads to beach erosion and a thinning of the barrier. Usually, this condition makes the barrier more susceptible to storm breaching and tidal inlet formation. As speculated earlier, such a situation may have facilitated the recent breaching of Monomoy Island in 1978 and Nauset Beach in 1987.

Along the Buzzards Bay coast in Slocum River Embayment a spit with associated tidal inlet was formed and subsequently destroyed in a period of less than 50 years. The construction of the spit system that formed the inlet and its later destruction were a consequence of a period of sand abundance followed by sediment starvation (FitzGerald et al., 1986; Fig. 15). Before 1941 the inner embayment was open and a channel existed along Deepwater

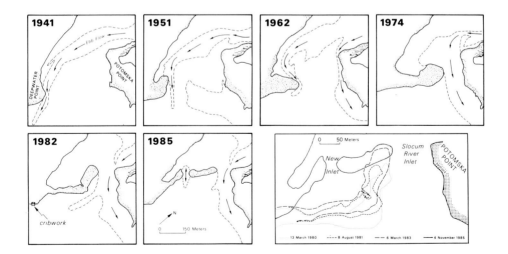

Figure 15. The construction and destruction of Slocum Spit as determined from vertical aerial photographs and field studies (from FitzGerald et al., 1986).

Point. Between 1941 and 1951 a spit began forming at Deepwater Point and accreted eastward across the bay. By 1974 the spit had deflected the main channel to a position along Potomska Point producing an inlet approximately 100 m wide. After the mid-1970's the spit began to erode and storm overwashing caused a landward migration of the barrier and onshore displacement of the inlet throat. The spit was breached in November 1984 during a spring tide and second inlet was formed adjacent to Deepwater point (Fig. 16). Since that time, the barrier continued to migrate onshore until it was transformed into an intertidal bar 50 m landward of its 1985 position (Mello, pers. comm.). As this process proceeds, the inner bay will return to an open water embayment and the tidal inlet will disappear.

The sedimentation history within Slocum River embayment suggests that a discrete supply of sand was responsible for forming the spit system and Slocum River Inlet. FitzGerald et al. (1987) speculate that the 1938 Hurricane transported sand from the Allens Pond barriers into the embayment (Fig. 14). Once the sediment was inside the bay, low wave energy gradually moved the sand toward Deepwater Point. A series of partially buried beach ridges in the marsh system landward of the shoreward migrating bar suggests that spit and tidal inlet formation process has occurred several times in the past in this embayment.

Closure And Openings of Tidal Inlets

Numerous tidal inlets have opened and closed along the Massachusetts coast in historic times and more would have closed if dredging projects had not been undertaken and engineering structures had not been constructed. A partial list of inlet openings and closures is given in Tables 1 and 3, respectively. To illustrate the conditions that led to inlet openings and closure several case studies are presented below.

Shirley Gut

Prior to the mid 1930's, Shirley Gut was a tidal inlet that separated Deer Island and Point Shirley along the northeast shore of Boston Harbor (Fig. 17A). The earliest surveys and charts of this region indicate that the inlet was 146 m wide

Figure 16. Oblique aerial photographs of Slocum River Inlet in: A. 1983 and B. 1985.

Table 3. Inlets which have closed along the Massachusetts coast
(partial list). Location shown in Figure 1.

INLET	LOCATION	HISTORY
Shirley Cut	Winthrop	Closed in 1934-36
South River	Marshfield	Closed During a Northeast Storm in 1898
Scusset Mills	Sagamore	Closed when the Cape Cod Canal was Built
East Harbor	Provincetown	Closed in 1869 Forming Pilgrim Lake
Katama Bay	Martha's Vineyard	Closed in 1869, 1915, 1934, 1969

at the inlet throat in 1860 and 10.7 m deep in 1847, shoaling to 7.2 m by 1861 (Nichols, 1949; Figs. 17B and D). During the next 70 to 75 years, the inlet narrowed and shoaled and by 1934 the inlet channel was barely subtidal and only 25 m wide at mean high water. A narrow isthmus (30 m wide) joined Point Shirley and Deer Island in 1936 (Nichols, 1949). Eventually the isthmus was broadened and filled to provide better access to facilities on Deer Island.

The inlet closed sometime between 1934 and 1936 and it has been suggested that storm processes contributed greatly to filling the inlet (FitzGerald, 1980). The short-term stability of the channel cross section during various periods of the inlet's history (1861 to 1869, Fig. 17B) suggests that under normal wave conditions the inlet was probably stable and tidal scour was sufficient to remove sediment dumped into the channel by longshore sediment transport. However, during storms stronger wave energies would have substantially increased the transport of sand and gravel from along Deer Island and Point Shirley toward the inlet. During the Blizzard of 1978 this region was the site of considerable deposition, including large gravel washovers greater than 1 m thick (FitzGerald, 1981). Although some of the sediment dumped into the inlet during storms would have been removed by increased currents caused by the accompanying storm surge, much of the sediment probably remained. The reason for this is that during storms most of the increased flow into and out of northern Boston Harbor was accommodated through President Roads channel. Thus, at Shirley Gut a disequilibrium was established during storms between the volume of sediment delivered to the inlet and the quantity of sediment scoured by tidal currents; this condition led to closure of the inlet.

Katama Bay Inlet

Katama Bay Inlet on the southeastern shore of Martha's Vineyard (Fig. 11) has opened and closed numerous times during the past 150 years (Ogden, 1974). As seen in Figure 18, breaching of Norton Point spit normally occurs in the middle of the barrier and is commonly associated with major storms

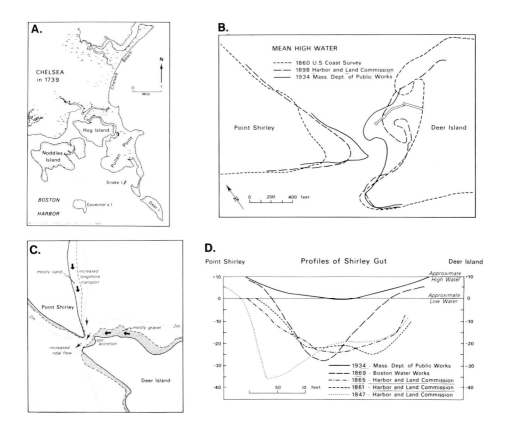

Figure 17. Historical account of the closure of Shirley Gut at Boston Harbor: A. early map of the region in 1739, B. and D. morphological changes of the inlet channel (from Nichols, 1949), and C. storm processes responsible for closing the inlet (from FitzGerald, 1980).

(Ogden, 1974). The opening which formed in 1886 was caused by a severe January northeast storm (Whiting, 1887). Breachings of the barrier, in approximately the same location, were produced by the 1938 Hurricane and Hurricane Carol in 1954. The February Blizzard of 1978 opened a small breach along the western part of the barrier, but this incipient breach immediately closed (Hanson and Forrester, 1978). Man-made cuts through the barrier were attempted in 1871, 1873, 1919 and 1921; only the last of these was successful (Ogden, 1974). After an inlet is cut, it migrates to the east and eventually closes as the spit attaches to Wasque Point. Inlet closure occurred in this manner in 1869, 1915, 1934 and 1969; since 1969 it has remained closed.

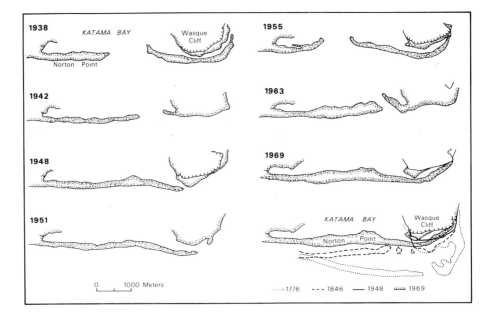

Figure 18. Shoreline changes at Katama Bay, Martha's Vineyard (from Ogden, 1974).

The instability of Katama Bay Inlet is related to a number of factors including: a strong easterly longshore transport system, a small tidal range (TR = .8 m), the shallowness of the southern end of Katama Bay that includes numerous intertidal shoals, and a northern deep channel opening to Katama Bay at Chappaquiddick Point (Fig. 11). The easterly movement of sand along the southern shoreline of Martha's Vineyard produces an eastward extension of the Norton Point spit and an easterly migration of Katama Bay Inlet. As the inlet moves farther to the east, the main tidal channel in the bay elongates and flow at the inlet becomes less efficient (cf., Keulegan, 1967). Also, repeated historical migrations of the inlet have produced numerous flood-tidal delta deposits which obstruct flow and provide an intertidal east-west barrier between the northern and southern portions of the bay. The most important factor which has led to the historical instability of the Katama Bay Inlet is the presence of the relatively deep inlet channel within Edgartown Harbor (Fig. 11). Most of the Katama Bay tidal prism is exchanged through this passage. If this opening did not exist, the ephemeral Katama Bay Inlet would be much larger and would probably remain open.

Allens Pond Inlet

The history of Allens Pond Inlet on the northwestern coast of Buzzards Bay is depicted over a 46-year period in Figure 19. During this time the inlet repeatedly migrated westward at an average rate of 100 m/yr, slowing considerably when it reached a far westerly position. The westward migration of the inlet is in opposition to the dominant easterly longshore transport direction and is caused by the configuration of the backbarrier main channel, which runs parallel to the eastern spit before turning southward at the inlet mouth (Fig. 19). Ebb flow in this channel is directed at the western inlet shoreline, causing erosion of the western channel bank in a manner similar to that of the cut bank side of a river meander. The westerly accreting spit is the point bar side of the meander bend and receives its sediment from sand that has been eroded at the inlet mouth and transported seaward to the eastward longshore transport system (FitzGerald et al., 1987). This mode of updrift inlet migration has been desoribed by Aubrey and Speer (1984) as one of the mechanisms reponsible for the northerly updrift migration of Nauset Inlet.

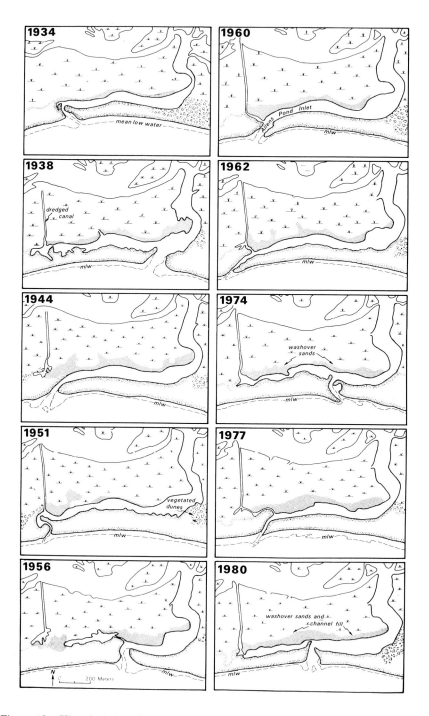

Figure 19.   Historical shoreline changes at Allens Pond Inlet. This inlet migrates updrift and periodically closes until a channel is cut at the eastern end of the spit (from FitzGerald et al., 1987).

As the inlet migrates to the west, flow in the elongating inlet channel becomes increasingly inefficient (Keulegan, 1967). When it reaches a far westerly position as it did in 1934, 1951, 1962, and 1977 (Fig. 19), the inlet narrows and eventually closes. Because the bay behind the inlet, Allens Pond, is a productive shellfishing area, the Town of South Dartmouth dredges a new inlet at the eastern end of the spit whenever this happens. Since 1980, an inlet has been cut through the spit in 1985 and again in 1989. When the inlet closed most recently, a rainy autumn resulted in the marsh being flooded by approximately 30 cm of water and salinity in the bay dropping to 7‰ (Mello, pers. comm.). Thus, to maintain an opening at Allens Pond Inlet, a new inlet must be cut every 5 to 10 years.

Briggs Marsh Pond and Quicksand Pond

Briggs Marsh and Quicksand Ponds are elongated bays situated on the northwestern shore of Buzzards Bay (Fig. 14). Although the ponds are actually located on the Rhode Island coast, the inlets that open and close along their shores are similar to the behavior of some inlets on the southern coasts of Cape Cod, Martha's Vineyard and Nantucket (Fig. 11) and thus, they are included in the discussion here. As explained earlier, the small potential tidal prisms of these ponds and the exposure of this coast to wave energy prohibit permanent inlets along their barrier shores. However, significant inflow of freshwater during late winter and early spring, which is derived from precipitation and melting snow and ice, produces outlets at these sites and other ponds along this coast (FitzGerald et al., 1987). The discharge channels are narrow (< 20 m) and form when the pond overtops the height of the barrier, usually at the site of a previous inlet. The depth of the outlet channel, which seldom exceeds mean low water, and overall dimensions of the cut through the barrier are dependent on the volume of water discharged and hydraulic head between the pond and ocean level. Because the barriers that front these ponds are composed of sand and gravel and are commonly porous, some water is continually discharged through the barriers themselves.

As the outlets become fully developed, tidal exchange between the ocean and bay establishes tidal inlet processes including the formation of recured spits, inlet migration, and the development of flood-tidal deltas (Greacen et al.,

1983; Fig. 20). Similar processes have been described at other tidal ponds on the Rhode Island coast (Boothroyd et al., 1985). Tidal inlets also develop at these ponds during intense storms such as the 1938 Hurricane and the Blizzard of 1978. Regardless of how the inlets formed, they are ephemeral features which close soon after they open, usually within several months. As seen in the maps of Quicksand and Briggs Marsh Ponds, the construction of flood-tidal deltas is an important process in impeding flow through the inlets and causing their closure (Fig. 20).

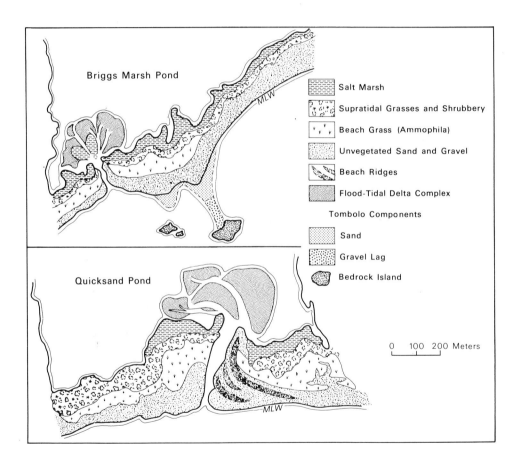

Figure 20. Sedimentary environments of the barriers fronting Briggs Marsh and Quicksand Ponds. Emphemeral inlets form along these shores due to storms and the discharge of freshwater. Note the development of flood-tidal deltas which restrict flow through the inlets (from FitzGerald et al., 1987 and modified after Greacen et al., 1983).

Scusset Mill Creek Inlet

Prior to the construction of the Cape Cod Canal in the early 1900's, Scusset Mill Creek emptied into Cape Cod Bay through Scusset Beach (Fig. 1). However, after the canal was completed and jetties were constructed to stabilize the entrance channel, the inlet closed. In digging the canal the headward portion of Scusset Mill Creek was connected to the waterway and thus its tidal prism was simply diverted to the canal. The accumulation of sand updrift of the north jetty caused a 200 m progradation of Scusset Beach and all traces of Scusset Mill Creek Inlet disappeared.

New Inlet, Scituate

New Inlet is located at the confluence of the North and South Rivers along the Massachusetts south shore (Fig. 21). The present inlet is anchored next to Fourth Cliff, one of several drumlins along this section of shoreline. In its early history North River was known for its shipbuilding and an American pirate who preyed on unsuspecting merchant ships journeying to and from Boston during the 1700's. At that time the entrance to the South and North rivers was 5.5 km south of Fourth Cliff along the southern end of Humarock Beach (Fig. 21). The inlet remained in that location until 1898, when a winter storm breached the sandy barrier that joined Third and Fourth Cliffs and New Inlet was formed. After the original breach, the inlet migrated slightly to the south and stabilized against Fourth Cliff. Much of the sand that originally comprised the barrier beach between Third and Fourth Cliffs has been reworked onshore in the form of two transgressive spits that extend southwest from Third Cliff and northeast from Fourth Cliff (Fig. 21). New Inlet is presently bordered on the north by a wide, low beach and intertidal shoal complex (Fig. 2). The stability of New Inlet will depend on the longevity of Fourth Cliff and the future of Humarock Beach. The thin parts of Humarock Beach and its overall erosional history have left this barrier highly susceptible to breaching, especially the neck region where South River impinges along the backside of the barrier (Fig. 21).

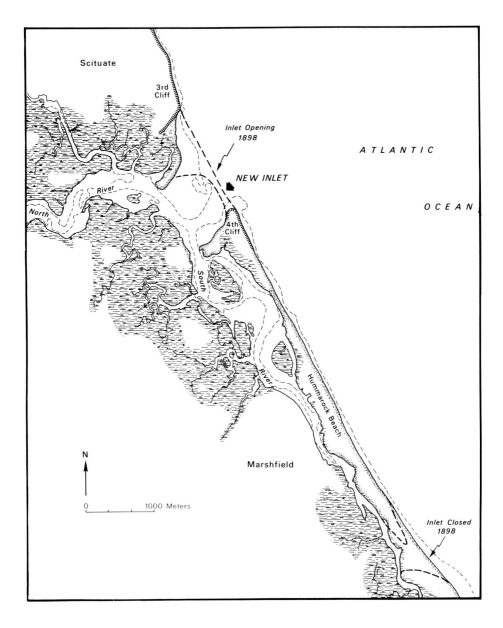

Figure 21. The northeast storm of 1898 caused the breaching of the barrier beach between Third and Fourth Cliffs. As New Inlet became established, the old inlet at the southern end of Hummarock Beach closed due to its greatly reduced tidal prism.

Jettied Inlets

There are more than 20 jettied inlets in Massachusetts, most of them concentrated along the microtidal shorelines of Nantucket Sound, Vineyard Sound, and Buzzards Bay (Table 1). Jetties are constructed to improve navigation through tidal inlets and depending upon their engineering design, they may stabilize a migrating inlet or its main channel, prevent or restrict the longshore transport of sand from entering the inlet channel, provide protection from wave action, and constrict or direct the ebb flow of an inlet to scour a deeper entrance channel. Jetty projects in Massachusetts have had mixed results, but in few instances have they provided the ultimate solution to navigation problems and in many cases they have created additional sedimentation and erosional problems. The effects of jetties along the Massachusetts shoreline are discussed using several inlet case histories.

Merrimack River Inlet

Tidal inlets situated along coasts where the longshore transport rate is comparable or greater than the potential sediment discharge in the inlet channel, often migrate or have main ebb channels that are deflected in a downdrift direction (Bruun and Gerritsen, 1959; Oertel, 1975). This condition was prevalent at the mouth of the Merrimack where the inlet had a history of southerly migration followed by a breaching back to the north prior to being stabilized by jetties in 1881 (Fig. 22; Hubbard, 1975; U.S. Army Corps of Engr., 1976). Since 1881, the jetties have been gradually extended so that now the north jetty is 1,250 m long and the south jetty is 745 m long. The shoreline has prograded on both sides of the inlet due to trapping of littoral drift moving south along Salisbury Beach during northeast storms and moving north along Plum Island by waves that refract around the ebb-tidal delta (Hubbard, 1975). After the initial progradation of the beach, the northern shoreline stabilized as further sand accumulation against the jetty was removed by storm waves and transported over the jetty into the inlet channel (Hubbard, 1973).

The northern end of Plum Island has a peculiar shape, in that a basin exists in the middle of the barrier (Fig. 22). Bathymetric surveys indicate that the basin

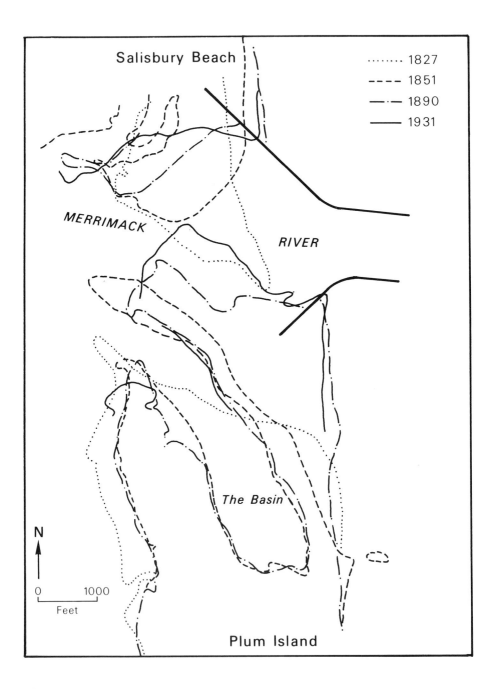

Figure 22. Shoreline changes at Merrimack River Inlet (from Nichols, 1941).

is almost 9 m deep (at mean low water). Although some of the depth may be attributed to dredging, it appears as though the basin represents a former position of the Merrimack River Inlet channel that existed sometime between 1827 and 1851. Nichols (1964) suggested that the basin formed due to a retrogration of northern Plum Island, followed by development of a spit along northern Plum Island which built northward. This scenario seems unlikely because the dominant longshore transport direction is to the south in this area (Hubbard, 1975), and thus the spit would have been accreting in opposition to this.

It appears more likely that ebb-tidal delta breaching took place at this inlet (Fig. 23) (cf., FitzGerald, 1988). In this model, the southerly longshore transport of sediment caused a preferential addition of sand to the northern, updrift side of the ebb-tidal delta (Fig. 23, Stage 1). In turn, this accumulation produced a southerly deflection of the main ebb channel and erosion of the northeast end of Plum Island (Fig. 23, Stage 2). In this configuration the circuitous pathway of the main channel produces inefficient flow through the inlet, resulting in a new channel being breached through the ebb-tidal delta sometime before 1851 (Fig. 23, Stage 3). The sand that had been on the updrift side of the ebb delta migrated onshore forming a large arcuate bar (U.S. Army Corps of Engr., 1976). In time the bar attached to Plum Island at its southern end and further sand supply from the ebb delta built the bar above mean high water (Fig. 23, Stage 4). This mechanism of inlet sediment bypassing and basin development along the downdrift inlet shoreline also has been identified on the South Carolina coast at Price and Capers Inlets (FitzGerald, 1982). The major effects of the Merrimack River Inlet jetty project have been a stabilization of the inlet channel and initial accretion of the shoreline on both sides of the inlet. However, subsequent shoaling at the inlet has required several emergency dredging projects by the U.S. Army Corps of Engineers. Erosion of the shoreline south of the jetties has resulted in beach nourishment projects and the construction of seven groins by the Commonwealth of Massachusetts.

Bass River Inlet

Jetties interrupt the natural mechanisms whereby tidal inlets bypass sand and in doing so, cause a progradation of the updrift shoreline while the sand-

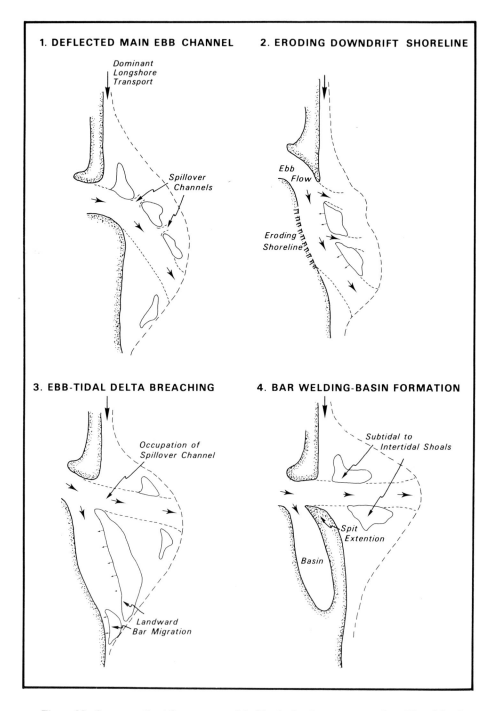

Figure 23. Conceptualized four stage model of basin development at northern Plum Island.

starved downdrift shoreline retreats. Bass River Inlet, located along the southern Cape Cod shoreline, is a good example of this condition (Figs. 2 and 24). A double jetty system, which was constructed at the inlet between 1866 and 1889, interrupts the easterly longshore transport system that dominates this coast. A net longshore transport rate of 4,400 m³/yr was calculated for this shoreline from the volume of sand that was trapped by the west jetty (177,000 m³) during a 40-year period from 1938 to 1978 (Slechta and FitzGerald, 1982). This is a minimum estimate because it does not account for the sand that bypassed the jetty during this time. Even at this relatively low transport rate, sand deposition along the Yarmouth shoreline necessitated a 30 m extension of the west jetty in the early 1950's (Fig. 24).

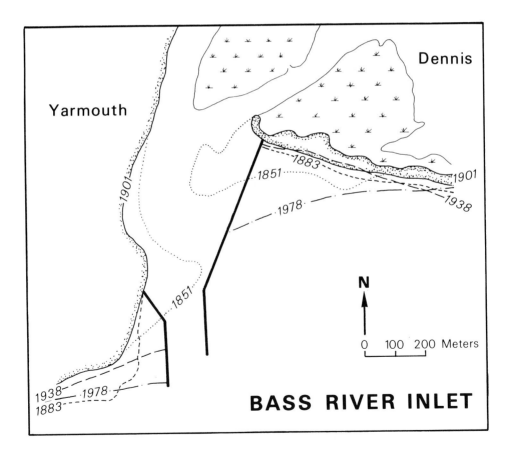

Figure. 24. Shoreline changes at Bass River Inlet as the result of jetty construction (from Slechta and FitzGerald, 1982).

In conjunction with the construction of the jetties in the 1800's, the interior of Bass River Inlet was dredged to provide a straight seaward path for the main channel. The jetties were designed to stabilize the entrance channel and alleviate shoaling problems. However, the small tidal prism of the inlet, coupled with periodic leakage of sand around the west jetty and transport of sand over the east jetty, have caused continuous shoaling problems. The shallowness of the offshore in this region, which is less than 2 m deep (mlw) at a distance of 1.5 km from the shoreline, exacerbates the shoaling problems seaward of the jetties (Slechta and FitzGerald, 1982).

Green Harbor Inlet

One consideration when constructing jetties is their orientation with respect to the approach of storm waves and the dominant wave swell of the region. For instance, at Wells Inlet along the coast of southern Maine the double jetty system is oriented into the direction of dominant wave approach (Byrne and Zeigler, 1977) making navigation through the jettied channel during storms and even periods of large swell extremely treacherous. In addition, wave action in the channel has been shown to contribute to the landward transport of sand through the inlet which has caused shoaling problems in the harborage area (Mariano and FitzGerald, 1991).

Green Harbor Inlet, located along the Massachusetts South Shore (Figs. 1 and 2), has double jetties oriented to the southeast away from the dominant northeast storm waves (Weishar and Aubrey, 1988). Current measurements taken at three sites across the channel at the mouth of the jetties on 31 August and 2 September 1981 indicate that the tidal currents are weak (max. vel. < 40 cm/sec), and that ebb velocities are consistently stronger than flood currents (FitzGerald, unpub. data). Based on tidal current measurements alone, the data suggest that the inlet should not import sediment. However, the entrance channel has had a long history of shoaling problems and has been dredged on several occasions. Interestingly, the material that accumulates in the entrance channel consists of sand and gravel. Some of this sediment is transported over the northern jetty during major storms and evidence of this process can be seen in Figure 2. However, most of the sediment moves landward into the inlet through the mouth of the jetties (Weishar and Aubrey,

1988). It would appear that the situation at Green Harbor Inlet is comparable to Wells Inlet, whereby shoaling waves in the main channel generate instantaneous, landward-directed currents that augment flood currents while retarding ebb currents. This process, especially during storms, would account for sediment transport into the inlet.

Construction of dikes, roadways, and retaining walls in the backbarrier region of Green Harbor have contributed to weak currents in the inlet channel. These structures cordoned off portions of the bay, marsh system, and tidal flats, thereby decreasing the volume of water entering and leaving the inlet. The tidal prism was reduced as a consequence of these modifications, so the size of the equilibrium entrance channel decreased.

Pamet River Inlet

Prior to stabilization, Pamet River Inlet in northern Cape Cod Bay had a history of northerly migration with periodic breachings of the spit to the south (Figs. 2 and 25). A new inlet was cut through the spit in 1919 and stabilized

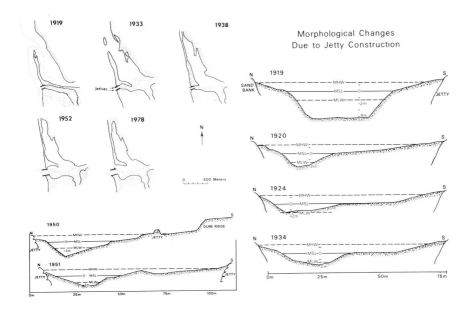

Figure 25. Morphological changes at Pamet River Inlet due to jetty construction (from FitzGerald and Levin, 1981)

with two stone jetties. The new inlet was 70 m wide and dredged to a depth of 4 m below mean low water. When the inlet was dredged, the spoil was placed landward of the inlet in the form of a dike which partitioned the backbarrier into two bays (Fig. 25, Year 1919). The jettied inlet quickly shoaled and it was not until the old inlet was closed and the dike severed that the combined tidal prisms of the two bays deepened the channel to 3.1 m by 1950 (FitzGerald and Levin, 1981).

In 1951 the jetties and channel were widened to 100 m to accomodate increased traffic through the inlet. The equilibrium channel cross section responded by shoaling more than a meter (Fig. 25, Years 1950 and 1951). At the same time the jetties were widened, they were also lengthened 50 m. In this configuration the south jetty was an efficient sediment trap of the northerly longshore transport, resulting in a 50 m progradation of its shoreline and a retreat of the downdrift northern shoreline. This retreat continued such that the north jetty was separated from the dune system and presently 10 to 20% of the tidal prism leaks between the north jetty and the adjacent beach. On the south side of the inlet, the beach has built to the end of the jetty and sand is readily transported into the inlet. Shoaling waves combined with flood currents move lobes of sand into the backbarrier region. Little of this sediment is removed by ebb tidal currents as evidenced by the beach ridges that are forming inside the inlet on the southern side of the bay (Fig. 2). Presently, access through the jettied channel can only be accomplished near high tide (FitzGerald and Levin, 1981).

The shoaling problems at Pamet River Inlet have resulted from poor jetty design and construction projects in the backbarrier. During the period from the mid-1800's to the early 1900's, the building of roads and railroad trestles has required portions of the marsh and bay to be diked. These dikes have isolated parts of the marsh in some areas and restricted tidal flow in others. As tidal strength was diminished, sedimentation in the backbarrier proceeded at a faster rate. These processes combined with the landward transport of sand into the bay have reduced the tidal prism and decreased the inlet equilibrium cross sectional area (FitzGerald and Levin, 1981).

Summary

Tidal inlets occur along the entire Massachusetts coast but are most common where the sand supply is most abundant, including the barrier coast north of Cape Ann and the sandy coasts of Cape Cod. Inlets are less common along the sand-starved coasts of Cape Ann and Buzzards Bay. Development of tidal inlets is closely related to barrier formation and constriction of an embayment. In Massachusetts, the constriction of embayments has occurred most commonly through spit accretion due to the presence of irregular coastlines (e.g., Sandy Neck and Nauset Spit), however, the formation of beach ridge barriers in front of bays (e.g., Horseneck Beach) is also an important process. Bays associated with Massachusetts inlets have developed in a variety of different settings. Drowned river valleys, groundwater sapping channels, kettles, and the enclosure of rocky and sandy irregular coasts are examples.

Tidal inlets in Massachusetts exhibit highly variable morphologies and sedimentation processes as a result of varying physical parameters including wave and tidal energies, sediment supply, and size of the backbarrier. Generally, inlets along mesotidal shorelines have well-developed ebb- and flood-tidal deltas and their backbarrier areas usually consist of marsh and tidal creeks with some open water areas near the inlet mouth. Plymouth Bay Inlet is an exception in that its backbarrier is dominated by intertidal flats. Inlets on microtidal shores commonly have poorly developed ebb-tidal deltas, poor-to-well developed flood-tidal deltas and backbarrier areas consisting of open bays with peripheral marshes (Hayes, 1975). The size of inlets is governed by bay area and tidal range.

Stability of unstructured tidal inlets along the Massachusetts shore is dependent on bay size, physical setting, and the erosional-depositional history of the associated barriers. Tidal inlets that have closed during the past 100 yrs generally are those along microtidal shores that had small bay areas or those where the tidal prism was rerouted through another inlet. Several barriers along the Buzzards Bay coast and the shores of Martha's Vineyard and Nantucket have ephemeral inlets which open during storms or when snow melt and precipitation cause the pond level to exceed the height of the barrier. Draining of the pond produces a cut through the barrier leading to inlet formation. Inlets that originate in this manner typically exist for a period of less than six months. Breaching of barriers and formation of semi-permanent tidal inlets have occurred predominantly along the outer coast of Cape Cod.

There are numerous jettied inlets on the Massachusetts coast; they are particularly prevalent along microtidal coasts where small bays and small ocean tidal ranges generate small tidal prisms. Jetties constructed at these sites disrupt the littoral transport system causing updrift accretion and downdrift erosion. In most cases, periodic dredging of the inlet channel is required to provide adequate navigation.

Acknowledgments

The collection and analyzes of data presented in this paper were supported by grants and contracts from Massachusetts Coastal Zone Management, U.S. Army Corps of Engineers, U.S. Minerals Management Service, Lloyd Center for Environmental Studies, and numerous individual towns. The arduous fieldwork of several graduate students, including Andrew Magee, Andrew Bakinowski, Jamie Greacen, Alan Saiz, Rapi Som, and Cynthia Thomlinson contributed to the material presented in this paper. Robert Chambers, Larry Nelson, and Kenneth Tobbin provided the auger core data presented for Herring River Inlet. Steve Goodbred and Wendy Quigley helped prepare the manuscript and Eliza McClennen of Boston University Cartographic Lab drafted the figures.

References

Aubrey, D.G., and A.G. Gaines, 1982. Recent evolution of an active barrier beach complex: Popponesset Beach, Cape Cod, MA. Technical Report # WHOI-82-3, Woods Hole, MA, 75.pp.

Aubrey, D.G. and P.E. Speer, 1984. Updrift migration of tidal inlets. *J. Geology,* v. 92, p. 531- 546.

Aubrey, Twichell, D.C., and S.L. Pfirman, 1982. Holocene sedimentation in the shallow nearshore zone off Nauset Inlet, Cape Cod, MA. *Mar. Geology,* v. 47, p. 243-259.

Belknap, D.F., B.G. Andersen, R.S. Anderson, W.A. Anderson, H.W. Borns, Jr., G.L. Jacobson, J.T. Kelley, R.C. Shipp, D.C. Smith, R. Stuckenrath, Jr., W.B. Thompson and D.A. Tyler, 1987. Late Quaternary sea-level changes in Maine. In: Nummedal, D., O.H. Pilkey and J.D. Howard, (eds.), *Sea-level Fluctuations and Coastal Evolution*, SEPM Spec. Pub. #41, p. 71-85.

Boothroyd, D.F., 1985. Tidal Inlets and Tidal Deltas. In: Davis, R.A. (ed.), *Coastal Sedimentary Environments*, Springer-Verlag, New York, p. 445-532.

Boothroyd, J.C., N.E. Friedrich and S.R. McGinn, 1985. Geology of microtidal lagoons, RI. *Mar. Geology,* v. 63, p. 35-76.

Boothroyd, J.C. and D.K. Hubbard, 1975. Genesis of bedforms in mesotidal estuaries. In: Cronin, L.E. (ed.), *Estuarine Research*, Academic Press, New York, v. 2, p. 217-234.

Bruun, P., 1967. Tidal inlet housekeeping. *J. of Hydraulic Divisions*, v. 93, p. 167-184.

Bruun, P., and F. Gerritsen, 1960. Natural bypassing of sand at coastal inlets. *J. of the Waterways and Harbors Div., Amer. Assoc. of Civ. Eng.,* v. 85, p. 75-107.

Byrne, P. and J.M. Zeigler, 1977. Coastal engineering study, Wells Harbor, ME. U.S. Army Corps of Engr., NE Division, Waltham, MA, 88 pp.

D'Amore, D.W., 1983. Hydrogeology and geomorphology of Great Sandford outwash plain, York County, ME with particular emphasis on the Branch Brook Watershed. Ph.D. Diss., Geology Dept., Boston Univ., Boston, MA, 147 pp.

Davis, R.A. and M.O. Hayes, 1984. What is a wave-dominated coast? *Mar. Geology,* v. 60, p. 313-329.

Davis, W.M., 1896. The outline of Cape Cod. American Academy of Arts and Sciences, Proceedings, v. 31, p. 303-332.

Edwards, G.B., 1988. Late Quaternary geology of northeastern Massachusetts and the Merrimack Embayment, western Gulf of Maine. Unpub. Masters Thesis, Geology Dept., Boston University, Boston, MA, 213 pp.

Fisher, J.J., 1987. Shoreline development of the glacial Cape Cod Coastline. In: FitzGerald, D.M., and P.S. Rosen, (eds.), *Glaciated Coasts,* Academic Press, New York, p. 280-307.

FitzGerald, D.M., 1976. Ebb-tidal delta of Price Inlet, SC: geomorphology, physical processes and associated inlet shoreline changes. In: Hayes, M.O. and T.W. Kana, (eds.), *Terr. Clastics Depos. Envir.,* Coastal Research Division, Dept. of Geol., Univ. of South Carolina, Columbia, SC, p. 158-171.

FitzGerald, D.M., 1981. Storm-generated sediment transport patterns along Winthrop and Yirrell Beaches, Winthrop, MA. *Northeast Geology,* v. 3, p. 202-211.

FitzGerald, D.M., 1980. Yirrell Beach Dynamics and Managgnent Guidelines. Tech. Rpt. #2, Coastal Envir. Res. Group, Geology Dept., Boston University, Boston, MA, 42 pp.

FitzGerald, D.M., 1982. Sediment bypassing at mixed energy inlets. *Amer. Soc. of Civil Engr.,* Proc. 18th Coastal Eng. Conf., p. 1094-1118.

FitzGerald, D.M., 1988. Shoreline erosional-depositional processes associated with tidal inlets. In: Aubrey, D.G. and L. Weishar, (eds.), *Hydrodynamics and Sediment Dynamics of Tidal Inlets,* Springer-Verlag, New York, p. 186-225.

FitzGerald, D.M., C.T. Baldwin, N.A. Ibrahim and D.R. Sands, 1987. Development of the northwestern Buzzards Bay Shoreline, MA. In: FitzGerald, D.M. and P.S. Rosen, (eds.), *Glaciated Coasts*, Academic Press, New York, p. 327-357.

FitzGerald, D.M. and D.R. Levin, 1981. Hydraulics morphology and sediment transport patterns at Pamet River Inlet, Truro, MA. *Northeastern Geology*, v. 3, p. 216-224.

FitzGerald, D.M., J.M. Lincoln, L.K. Fink and D. Caldwell, 1990. Morphodynamics of tidal inlet systems in Maine. In: *Maine Geological Survey Studies in Maine Geology*, Augusta, ME, v. 5, p. 1-30.

FitzGerald, D.M., A. Saiz, C.T. Baldwin and D.M. Bands, 1986. Spit breaching at Slocum River Inlet, Buzzards Bay, MA. *Shore and Beach*, v. 54, p. 11-17.

Friedrichs, C.T., D.G. Aubrey, G.S. Giese and P.E. Speer. Hydrodynamical modeling of a multiple-inlet estuary/barrier system: Insight into tidal inlet formation and stability. In: D.G. Aubrey and G.S. Giese (eds.), *Formation and Evolution of Multiple Inlet Systems, Coastal and Estuarine Studies*, (this volume).

Gatto, L.W., 1978. Shoreline changes along the outer shore of Cape Cod from Long Point to Monomoy Point. CRREL Rpt. #78-17, U.S. Army Corps of Engr., Hanover, NH, 49 pp.

Giese, G.S., 1988. Cyclic behavior of the tidal inlet at Nauset Beach, Chatham, Massachusetts. In: Aubrey, D.G. and L. Weishar, (eds.), *Hydrodynamics and Sediment Dynamics of Tidal Inlets*, Springer-Verlag, New York, p. 269-283.

Greacen, J., J. Rahilly and C. Tomlinson, 1983. An examination of coastal variation, Sakonnet Point, RI to New Bedford, MA. Unpub. Rpt. Coastal Environmental Research Group, Dept. of Geology, Boston Univ., Boston, MA, 17 pp.

Halsey, S.D., 1979. New model of barrier island formation. In: Leatherman, S.P. (ed.), *Barrier Islands from the Gulf of St. Lawrence to the Gulf of Mexico*, New York, Academic Press, p. 185-210.

Hanson, L.S. and V.S. Forrester, 1978. Coastal Guide to Martha's Vineyard. Geology Dept., Boston University, Boston, MA, 84 pp.

Hayes, M.O., 1975. Morphology of sand accumulation in estuaries: an introduction to the symposium. In: Cronin, L.E. (ed.), *Estuarine Research*, New York, Academic Press, v. 2, p. 1-22.

Hayes, M.O., 1979. Barrier island morphology as a function of tidal and wave regime. In: Leatherman, S.P. (ed.,), *Barrier Islands from the Gulf of St. Lawrence to the Gulf of Mexico*, New York, Academic Press, p. 1-28.

Hayes, M.O., E.H. Owens, D.K. Hubbard and R.W. Abele, 1973. The investigation of forms and processes in the coastal zone. In: Coates, D.R. (ed.), *Coastal Geomorphology*, Publications in Geomorphology, Binghampton, NY, p. 11-41.

Hill, M.C., D.M. FitzGerald and C.T. Baldwin, 1990. Development and assessment of sand resources in Plymouth Bay. U.S. Minerals Management Service Proc., Austin, TX, p. 182-193.

Hill, M.C. and D.M. FitzGerald. Evolution and Holocene stratigraphy of Plymouth Bay, MA. In: Fletcher, C.H. and J.F. Wehmiller (eds.), *Quaternary Coastal Systems of the United States*, SEPM Spec. Pub., (in press).

Hine, A.C., 1975. Bedform distribution and migration patterns on tidal deltas in the Chatham Harbor Estuary, Cape Cod, MA. In: Cronin, L.E. (ed.), *Estuarine Research*, New York, Academic Press, p. 235-252.

Hine, A.C., S.S. Snyder and A.C. Neumann, 1979. Coastal plain and inner shelf structure, stratigraphy and geologic history: Bougue Banks area, NC. Tech. Rpt. North Carolina Sci. Technology Comm., Chapel Hill.

Hubbard, D.K., 1973. Morphology and hydrodynamics of Merrimack Inlet, Massachusetts, Part I. Final Rpt. for Contract # 72-72-C-0032, Coastal Engr. Res. Center, 154 pp.

Hubbard, D.K., 1975. Morphology and hydrodynamics of the Merrimack River ebb-tidal delta. In: Cronin, L.E. (ed.), *Estuarine Research,* New York, Academic Press, v. 2, p. 253-266.

Ibrahim, N.A., 1986. Sedimentological and morphological evolution of a coarse-grained regressive barrier beach, Horseneck Beach, MA. Unpub. Masters Thesis, Dept. of Geology, Boston University, Boston, MA, 196 pp.

Inman, D.L. and J.D. Frautschy, 1965. Littoral processes and the development of shorelines. Coastal Engr. Spec. Conf., Santa Barbara, CA, p. 511-536.

Jarrett, J.T., 1976. Tidal prism-inlet area relationships. GITI Rpt. 33, U.S. Army Corps of Engr. Res. Center, Waterways Exp. Station, Vicksburg, MS, 55 pp.

Jensen, R.E., 1983. Atlantic coast hindcasting shallow water significant wave information. Waterways Experiment Station, Vicksburg, MS, 75 pp.

Keulegan, G.H., 1967. Tidal flow in entrances: water level fluctuations of basins in communications with seas. Tech. Bull. #14, Comm. on Tidal Hydraulics, U.S. Army Corps of Engr., Vicksburg, MS, 100 pp.

Larson, G.J., 1982. Nonsynchronous retreat of ice lobes from southeastern Massachusetts. In: Larson, G.J. and B.D. Stone, (eds.), *Late Wisconsinan of New England,* Kendall/ Hunt, Dubuque, Iowa, p. 101-115.

Lucke, J.B., 1934. A theory of the evolution of lagoon deposits on shorelines of emergence. *J. Geology,* v. 42, p. 561-584.

Magee, A.D. and D.M. FitzGerald, 1980. Investigation of the shoaling problems at Westport River Inlet and sedimentation processes at Horseneck and East Horseneck Beaches. Tech. Rpt. #3, Coastal Envir. Res. Group, Dept. of Geology, Boston University, Boston, MA, 118 pp.

Mariano, C.G. and D.M. FitzGerald, 1989. Sediment transport patterns and hydraulics at Wells Inlet, Maine. Tech, Rpt. #12, Coastal Envir. Res. Group, Dept. of Geology, Boston University, Boston, MA, 143 pp.

McIntire, G.W. and J.P. Morgan, 1964. Recent geomorphic history of Plum Island, MA and adjacent coasts. Coastal Series #8, Louisiana State Univ. Press, Baton Rouge, LA, 44 pp.

Morton, R.A. and A.C. Donaldson, 1973. Sediment distribution and evolution of tidal deltas along a tide-dominated shoreline, Wachapreague, Virginia. *Sed. Geol.,* v. 10, p. 285-299.

Nichols, R.L., 1941. The geology of Plum Island, Castle Neck, and Great Neck. In: Chute, N.E. and R.L. Nichols, (eds.), *The Geology of the Coast of Northeastern Massachusetts,* Mass. Dept. of Public Works, U.S. Geol. Surv. Coop. Proj., Bull #7, 48 pp.

Nichols, R.L., 1949. Recent shoreline changes at Shirley Gut, Boston Harbor. *J. Geology,* v. 57, p. 84-89.

Nichols, R.L., 1964. Northeastern Massachusetts geomorphology- Trip A. In: Skehan, S.J., (ed.), *Guidebook to Fieldtrips in the Boston Area,* NEIGC, p. 3-40.

National Oceanic and Atmospheric Administration, 1989. *Tide Tables, East Coast of North and South America.* Rockville, MD, 287 pp.

Nummedal, D. and I. Fischer, 1978. Process-response models for depositional shorelines: the German and Georgia Bights. *Amer. Soc. of Civil Eng.,* Proc. 16th Coastal Eng. Conf., p. 1215-1231.

O'Brien, M.P., 1931. Estuary tidal prisms related to entrance areas. *Civil Engr.,* v. 1, p. 738-739.

O'Brien, M.P., 1969. Equilibrium flow areas of inlets on sandy coasts. *Jour. of Waterways, Harbors, and Coastal Engrs.,* ASCE, v. 95, p. 43-55.

O'Brien, M.P. and R.G. Dean, 1972. Hydraulics and sedimentary stability of coastal inlets. Proc. of the 13th Annual Coastal Engr. Conf., Vancouver, BC, Canada, p. 761-780.

Oertel, G.F., 1975. Ebb-tidal deltas of Georgia estuaries. In: Cronin, L.E. (ed.), *Estuarine Research,* New York, Academic Press, p. 267-276.

Oertel, G.F., 1979. Barrier island development during the Holocene recession Southeastern United States. In: Leatherman, S.P. (ed.), *Barrier Islands from the Gulf of St. lawrence to the Gulf of Mexico,* New York, Academic Press, p. 273-290.

Ogden, J.G., 1974. Shoreline changes along the southeastern coast of Martha's Vineyard, MA for the past 200 years. *Quaternary Res.,* v. 4, p. 496-508.

Oldale, R.N. and R.A. Barlow, 1986. Geologic map of Cape Cod and the Islands, Massachusetts. U.S. Geological Survey, Woods Hole, MA.

Oldale, R.N. and C.J. O'Hara, 1980. New radiocarbon dates from the inner continental shelf of southeastern Massachusetts and a local sea-level curve for the past 12,000 years. *Geology,* v. 8, p. 102-106.

Penland, S., R. Boyd and J.R. Suter, 1988. Transgressive depositional systems of the Mississippi Delta plain: a model for barrier shoreline and shelf sand development. *J. Sed. Petrology,* v. 58, p. 932-949.

Pierce, J.W., 1976. Tidal inlets and washover fans. *J. Geology,* v. 78, p. 230-234.

Redfield, A.C., 1967. Ontogeny of a salt marsh. In: Lauff, G.H., (ed.), *Estuaries,* AAAS, Washington, DC, p. 108-114.

Rhodes, E.G., 1973. Pleistocene-Holocene sediments interpreted by seismic refraction and wash-bore sampling, Plum Island-Castle Neck, MA. Tech. Mem. #40, 75 pp.

Slechta, M.W. and D.M. FitzGerald, 1982. Sedimentation patterns and processes at Bass River Inlet. Tech. Rpt. #6, Coastal Envir. Res. Group, Dept. of Geology, Boston University, Boston, MA, 154 pp.

Som, R.M., 1990. Stratigraphy of the Castle Neck-Essex Bay marsh system. Masters Thesis, Geology Dept., Boston Univ., Boston, MA.

Stone, B.D. and J.D. Peper, 1982. Topographic control of the deglaciation of eastern Massachusetts: ice lobation and the marine incursion. In: Larson, G.J. and B.D. Stone, (eds.). *Late Wisconsinian of New England,* Kendall/Hunt, Dubuque, Iowa, p. 145-166.

Thompson, E.F., 1977. Wave climate at selected locations along U.S. coasts. Tech. Rpt. No. 77-1, C.E.R.C., U.S. Army Corps of Engr., Washington, D.C., 364 pp.

U.S. Army Corps of Engineers, 1957. Saco Maine beach erosion control study. House Document #32, 85th Congress, 37 pp.

U.S. Amy Corps of Engineers, 1976. Plum Island feasibilty report. Waltham, MA District Office, 25 pp.

Weishar, L.L. and D.G. Aubrey, 1988. Inlet hydraulics at Green Harbor, Marshfield, MA.
    U.S. Amy Corps of Engr. Misc. Paper CERC-88-10, Vicksburg, MS, 104 pp.
Whiting, H.L., 1887. Recent coastal changes at Martha's Vineyard, MA, Appendix 9, U.S.
    Coast and Geodetic Survey, Annual Report.

2

Morphodynamic Evolution of a Newly Formed Tidal Inlet

James T. Liu, Donald K. Stauble, Graham S. Giese and David G. Aubrey

Abstract

A unique opportunity to document and understand the processes of tidal inlet evolution and morphodynamic interactions presented itself with the breach of Nauset Spit across from the town of Chatham, Massachusetts. In the first twenty-eight months since its formation, the morphological evolution of the new inlet can be characterized into four categories: 1) Inlet mouth widening, caused by the concurrent retreats of the north and south spits that flank the inlet mouth; 2) Cyclical spit elongation, breaching, and terminal detachment; 3) The southward migration of the thalweg of the main channel and its associated shoals; and 4) Shoal growth. Six major shoals both seaward and landward of the inlet mouth developed between inlet formation and 1991. The growth of shoals and spit elongation indicate the trapping of littoral sediments within the inlet mouth area and the influx of ocean sediments entering the lagoon through the new inlet. The southward movement of the inlet channel is driven by its own channel configuration. As a result of the formation of this inlet, Chatham Harbor has evolved into two separate systems, each having its own hydrodynamic and morphological characteristics.

Formation and Evolution of Multiple Tidal Inlets
Coastal and Estuarine Studies, Volume 44, Pages 62-94

Introduction

Inlet formation by breaching of a narrow section in a barrier shoreline is common in the historical record of coastal areas. Initial inlet formation usually develops during storm conditions, when the storm surge raises the water level on either the oceanside or the bayside of the barrier. The increased water levels allow the storm-induced waves and swash processes to erode into the beach backshore and dunes. On narrow barrier shorelines, particularly where dunes are low and discontinuous, the swash often will breach between the dunes and overwash the barrier carrying water and sediment to the lagoonside of the barrier. If the storm duration is sufficiently long enough to allow repeated overwashing, particularly around high tide, storm waves can cut a breach of sufficient magnitude to create a channel. Once there is an opening to the back-barrier bay/lagoon, and if there is sufficient tidal forces to create ebb and flood flows through the breach, it may be sustained after the storm surge subsides. The interaction of the tidal and wave forces may subsequently maintain and modify the opening to create a persistent tidal inlet (see Friedrichs et al., this volume).

Once a tidal inlet comes into existence, it interrupts the wave-induced longshore sediment transport (FitzGerald, 1989), causing not only shoreline changes in the vicinity of the inlet, but also formation of shoals within the inlet mouth and on both landward and seaward sides of the inlet. The newly formed inlet acts as a conduit for sediment and water exchanges between the bay/lagoon and the ocean.

In the initial stage of inlet development following its formation, the tidal sand transport through the inlet has not obtained equilibrium with the littoral drift. Subsequently, the inlet is expected to show rapid changes in its adjacent shoreline configuration and shoal morphology as it interacts to the changing hydrodynamic conditions. These changes, however, lead the inlet towards an equilibrium state. This sequence of change reveals mechanisms by which the entire inlet/barrier-bay system responds to the formation of the tidal inlet.

On occasions when a new inlet forms in close proximity to another existing inlet and shares the same tidal prism, complex interactions among wave, tidal and littoral forces establish a new set of equilibrium conditions. Most literature regarding tidal inlet changes have focused on single inlets that have been in existence for a period of time, and have obtained some consistency in their configuration and shoal morphology. There has been little documentation on tidal inlet changes during the initial stage of their development. The present study provides a unique opportunity to examine the morphological changes of an evolving tidal inlet in the first 28 months following its formation.

Study Area

The study site, Chatham Harbor, is a bar-built estuary/lagoon located on the southeastern corner of Cape Code, Massachusetts (Fig. 1). The estuary is an elongate, coast-parallel body of well-mixed water, approximately 6 km long, 1 km wide, having a maximum depth of 7 m. The northern reach of the lagoon consists of the wider and shallow Pleasant Bay.

Before January 2, 1987, Chatham Harbor was sheltered from the Atlantic Ocean by a barrier spit called Nauset Beach. The opening of this estuary into the Atlantic Ocean was located at the southern tip of the spit through Chatham Bars Inlet (now referred to as South Channel, Fig. 2). Since the breach, the southern portion of the barrier spit has been cut off, forming a barrier island called South Beach. Monomoy Island was originally a 12.9 km long barrier island to the south and west of South Beach. During historic times, this island migrated westward toward Nantucket Sound (Giese et al., 1989) and is now bisected by a small unnamed inlet separating North and South Monomoy Islands (Fig. 2). An opening between Morris Island and North Monomoy Island, called West Channel, leads to Nantucket Sound.

Prevalent deepwater waves approach Nauset Beach from the E-NE quadrants (Wright and Brenninkmeyer, 1979). Wave observations immediately north of Nauset Beach indicate a wave climate having an equivalent significant wave height of 3 m observed in January and February (Aubrey et al., 1982). During the same period, southward mean flows having an average velocity of 6 cm/sec were also observed (Aubrey et al., 1982). As a result of the prevailing wave field, a net southward longshore sediment transport rate is estimated to be 5×10^5 cubic meters/year for the northern part of Nauset Beach

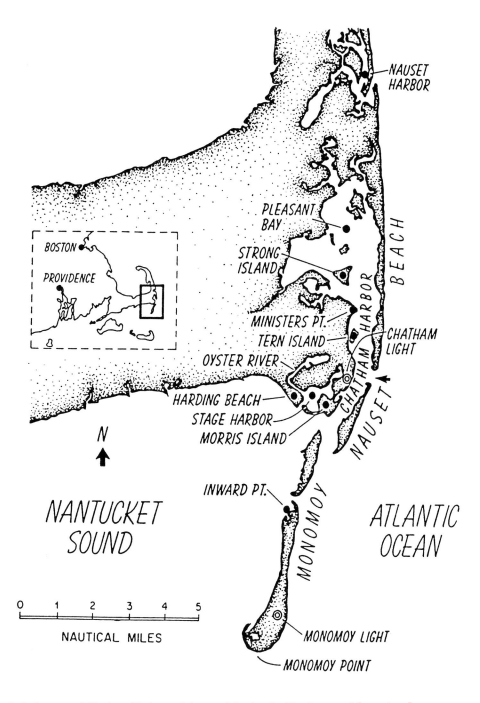

Figure 1. Index map of Chatham Harbor and the new inlet (marked by the arrow) formed on January 2, 1987. The shoreline drawn on this map (except the newly formed tidal inlet) was based on a map of 1980. The insert map shows the location of Chatham in Cape Cod, Massachusetts.

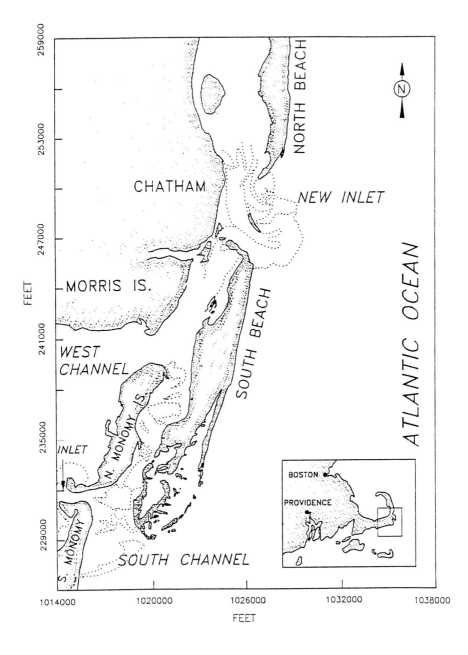

Figure 2. Updated map of Chatham Harbor estuary, with North Beach (spit), South Beach (barrier island) and North and South Monomoy Islands (reproduced from aerial photographs taken in September, 1988, and plotted on Massachusetts Grid). Along with the new inlet, the three previously existing inlets leading to the Atlantic Ocean and Nantucket Sound are now called South Channel, West Channel, and an unnamed inlet between N and S Monomoy Island.

(Cornillon, 1979). Due to their frequency and intensity, storms also play an important role in shaping the morphology of barrier spits in this area (Leatherman, 1979; McClennen, 1979; Aubrey and Speer, 1984).

On January 2, 1987, Nauset Beach was breached during a severe northeaster storm coinciding with a perigean high spring tide (Giese et al., this volume). The break-through first appeared as extensive washovers across the barrier beach when storm surge caused breaking waves to inundate the beach. When the storm subsided the next day, a meandering surge channel (approximately 5.5 m wide and 0.3 m deep) had been formed in the center of a washover fan in which the water continued to flow at both high and low tides. Aerial photographs taken a few days after the initial cut on Nauset Beach indicated that the transport of water had been predominantly bayward across the beach at high tides. In January and February 1987, Chatham experienced two more northeasters. By March, the breach had become 520 m wide.

The formation of the new inlet improved the efficiency of water exchange between the lagoon and ocean as indicated by the increased tidal range in Chatham Harbor after the breach. In April, 1988, near-bed measurements of the tidal current speed and tidal height within the channel throat of the new inlet showed that the average tidal range was 1.6 m. The maximum flood current speed exceeded 100 cm/sec, and the maximum ebb current speed exceeded 140 cm/sec (Fig. 3). On average, the maximum flood speed preceded the high water by about two hours, and the maximum ebb speed preceded the low water by about one hour and 45 min. This relationship indicates that the tide through the new inlet displayed mixed characteristics of both a progressive wave and a standing wave. The objective of this study is to provide initial, qualitative documentation on the morphological as well as bathymetric evolution of the newly formed inlet in Chatham.

Methods

This study included an analysis of a series of aerial photographs taken in a sequence starting shortly after the formation of the new inlet, to assess changes in shoreline and shoal morphology. A select set of sediment samples was collected and analyzed to characterize sediment grain-size distributions

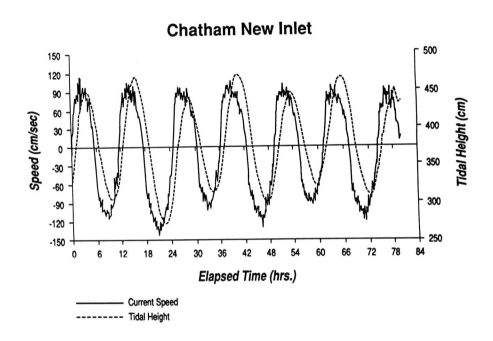

Figure 3. Tidal current speed and tidal height measured within the channel throat of the new inlet.

on a rapidly shoaling and evolving flood-tidal delta. Bathymetric survey
were done of the evolving channels and shoals around the new inlet mouth
Changes in shoreline position and inlet morphology were used to identify th
processes of inlet formation, morphodynamics and sediment transport path
ways active at the new inlet.

Aerial Photographs

Vertical aerial photographs of Nauset Beach and Chatham Harbor were taken at 4-month intervals beginning in May, 1987, through May, 1989. Except for the September 1987 set which was contracted by the National Park Service for another purpose, the time and altitude at which the photographs were taken were pre-determined to obtain maximum exposure of the intertidal shorelines and inter-to-subtidal shoals. A time sequence of 7 sets of aerial photographs has been obtained since the inlet formed (Table 1), along with a 1982 set of pre-breach conditions. Since several scales of photography were used in this analysis, a base map (1" = 1455') was constructed from a 1:24,000 USGS topographic map of Chatham. A Zoom Transfer scope was used to rectify and bring the various scales of the aerial photography to a common map scale. The shoreline shown on each set was digitized along the approximate low-tide waterline. In addition, the outlines of the inter- and subtidal shoals associated with the new inlet, such as the ebb-tidal delta, channel-margin bars and flood-tidal delta, were also digitized. The digitized photographs were then plotted on the Massachusetts Grid System (in feet) for further analysis. The reproduced maps were also carefully checked against one another for accuracy and precision, and were determined satisfactory for the qualitative purpose of this study.

Sediment Grain-Size Analysis

In September, 1988, eight surficial sediment samples were collected by hand near the time of low water from locations along a transect across the flood-tide delta (shoals) linking the spit on the South Beach to the inner shore of Chatham immediately south of the new inlet. These sediment samples represent recent depositional conditions of the previous falling tide. They were analyzed at the sediment laboratory of the Coastal Engineering Research Center of the Army Corps of Engineers, using a sonic sifter at quarter-phi sieve intervals. Grain-size frequency distributions and statistical data were then calculated using the Interactive Sediment Analysis Program (ISAP). Mean, sorting, skewness, and kurtosis were calculated using the method of moments.

Table 1. Time sequence of aerial photographs of Chatham Harbor.

Time	Scale	Remarks
October 1982	1:18000	Including portions of Monomoy Islands
May 1987	1:9000	
September 1987	1:8000	Excluding portions of inner shore of Chatham
January 1988	1:9000	
May 1988	1:18000	Including Monomoy Islands
September 1988	1:18000	Including Monomoy Islands
December 1988	1:18000	Including Monomoy Islands
May 1989	1:18000	Including Monomoy Islands

Bathymetric Survey

A bathymetric survey was conducted in April, 1988. A Del Norte microwave radar navigation system was used for positioning. Depth was measured using an Odom Echotrac fathometer on board a 19-foot Boston Whaler. The data were logged into a shipboard computer simultaneously with the position information using automated, integrated navigation software. The depths were later corrected for tidal elevations at the time of the survey. The survey covered the seaward edge of the terminal lobe of the ebb-tidal delta of the new inlet, part of the main channel around the throat of the new inlet, and channels of the two previously existing inlets.

Morphological Evolution of the New Inlet

Since the initial breach in January 1987, the new inlet has continued to grow and develop distinctive geomorphology. The interactions between tidal and wave forces during the 28-month study period have resulted in the evolution and migration of the newly formed tidal inlet, its associated shoals and channels, and adjacent shorelines. Specific details of the continued widening of the inlet and resulting complex changes in the shorelines of the north and south spits adjacent to the inlet are examined. Common inlet-associated shoals have formed and the evolution of these flood and ebb-tidal shoals are described.

Inlet-Mouth Widening and Spit Retreat

Maps of the low-tide shoreline (solid lines) and outlines of inter- to subtidal shoals and channels (dotted lines) around the new inlet (Figs. 4a-h) were produced from the aerial photo sets. The distance between the spit on North Beach and that on South Beach continued to increase between May, 1987, and May, 1989, allowing greater influence of ocean waves to impinge upon the interior of the estuary. This widening was accomplished by the southward retreat of the north end of South Beach (the south spit), and the contemporaneous northward retreat of the south end of North Beach (the north spit, Fig. 5). The shoreline changes along the axis of Nauset Beach were quantified by measuring the distances from the tips of the north and south spits, and from the north and south edge of the main channel at the throat, to the location where the breach first appeared (Fig. 6).

Within the trends of the progressive movements of the north and south spits, both spits showed short-term fluctuations in their orientations and lengths (Fig. 5). Casual observations indicated that the north spit retreated in a stepwise fashion that included southward lengthening followed by severe overwashing and subsequent detachment of the tip of the spit. The detached remnant spit soon became a subtidal shoal and disappeared. At times, after a shortening episode of terminal detachment, the tip of the north spit would grow southward, passing the previous position as exemplified by the positions of the north spit terminus in January, 1988 relative to September, 1987; September, 1988 relative to May, 1988; and May, 1989 relative to December, 1988 (Fig. 5). No seasonal pattern for the retreat of the north spit was established due to inadequate temporal coverage of the aerial photographs. However, the pattern of alternate spit truncation and elongation was evident.

Simultaneous with the progressive southward retreat of the north end of South Beach, the south spit also displayed cyclical elongation followed by breaching and terminal detachment. The detached terminus then became an intertidal island. Subsequently, a chain of the NW-SE oriented intertidal islands and flats has formed landward of the south spit as a result of the combination of both the cyclical E-W movements of the south spit and the progressive southward retreat of the north end of South Beach. Three such cycles during 1987 and 1988 have been observed, each having a period of approximately four months. Within each cycle, distinctive morphological stages of the spit were recognized (Ebert and Weidman 1989).

Although the frequency of aerial photographs of this study is out of phase with the spit cycle, different stages from different spit cycles were recorded. The south spit in September, 1987 (Fig. 4c) represents the initiating stage of a new cycle preceded by a breach. In May, 1988, the spit was probably in the middle-stage of rapid growth (Fig. 4e). September, 1988 (Fig. 4f) represents the late stage at which a breach is about to occur. The repetitive elongation of the south spit in its growth-and-breach cycles indicates a constant influx of littoral sediments into Chatham Harbor.

Shoal Development and Evolution

As the new inlet formed, a series of shoals became readily identifiable from the aerial photography. Six areas of shoals have been identified based on shoal configurations and previous inlet investigations (see for example Hayes, 1980, Boothroyd, 1985). The new inlet has developed many of the shoal morphologies commonly associated with tidal inlets. Figure 7 shows the generalized inter- to sub-tidal sand bodies labeled 1 through 8. Shoal features associated with the inlet mouth are the ebb-tidal delta (1) and flood-tidal deltas (5 and 6). The specific morphology of the flood-tidal deltas is

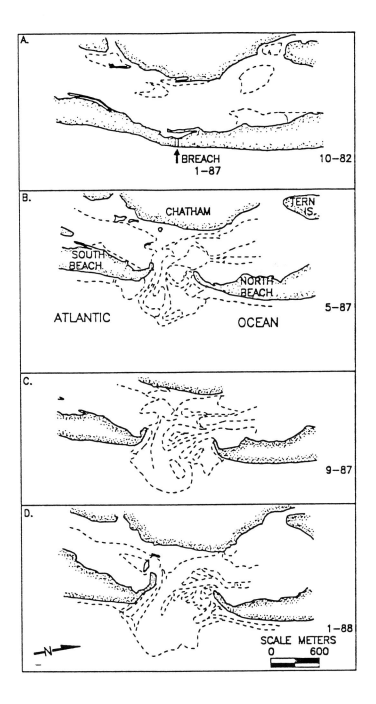

Figure 4. Development of shorelines (solid lines) and shoals (dotted lines) of the new inlet from aerial photography in: a) October, 1982; b) May, 1987; c) September, 1987; d) January, 1988; e) May, 1988; f) September, 1988; g) December, 1988; and h) May, 1989.

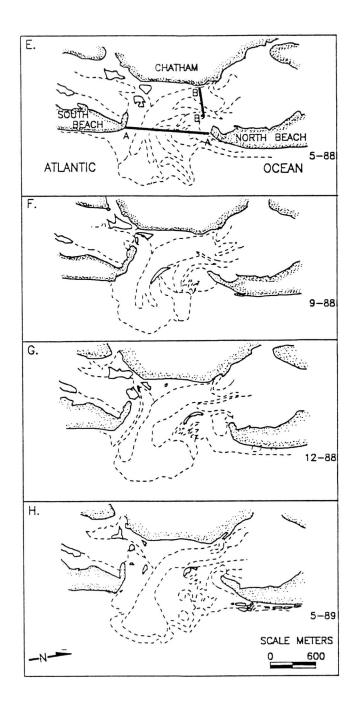

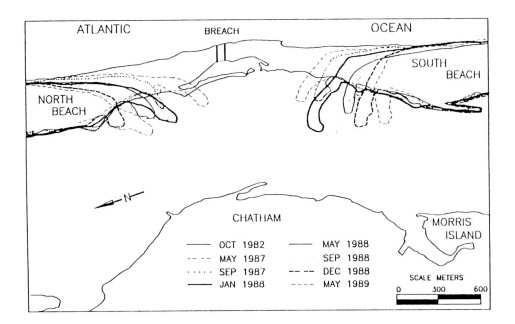

Figure 5. Sequential shoreline changes of the two spits flanking the new inlet, from May, 1987 to May, 1989. The pre-breach shoreline of October, 1982 is also plotted for comparison.

influenced by the long narrow estuary shape and close proximity of the Chatham mainland shore.

The ebb-tidal delta at the new inlet is the most prominent feature seaward of the inlet mouth. A large channel-margin swash platform (2) also developed on the updrift (north) side of the inlet channel throat. Landward of the inlet mouth, the flood-tidal shoal complexes are somewhat different from the classical inlet flood shoal development. The narrow, elongate, and coast-parallel Chatham Harbor constricts expansive flood shoal configuration. In addition, the presence of remnant shoals that predate the opening of the inlet also influence present shoal locations (Fig. 4a). Each of the north and south flood-tidal deltas is associated with remnant shoals (Fig. 7).

Two other remnant lagoon shoals also interact with the inlet morphodynamics. A linear shoal (7) trending parallel with Nauset Beach is visible on the

THROAT AND EBB CHANNEL WIDTH

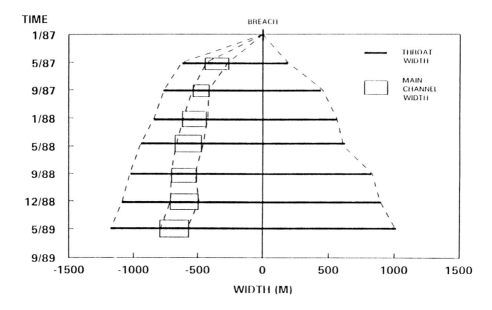

Figure 6. Temporal changes of the width of the new inlet mouth between north and south spits (solid lines) and the width and position of the main channel throat (rectangles) with respect to the location of the initial breach (the origin on the horizontal axis). The positive distance indicates the north side of the initial breach.

1982 aerial photography. This shoal seems to be related to tidal flow along the axis of the estuary and has been modified as the inlet develops. A sand flat (8) in front of the small boat harbor between Toms Neck and Morris Island is also present on the 1982 aerial photography and remains after the inlet opened.

As the new inlet continues to evolve both the ebb- and flood-tidal deltas grew in size. Table 2 gives the area of the shoals measured from the aerial photography. Numbers preceding each shoal or spit indicate the location on Figures 8a and 8b, which summarize these areal changes. Since subaqueous boundaries of each shoal were determined visually, this type of analysis is only indicative of sediment depositional changes that were visible from the aerial photography. No detailed bathymetry corresponding to the dates of aerial photography was available.

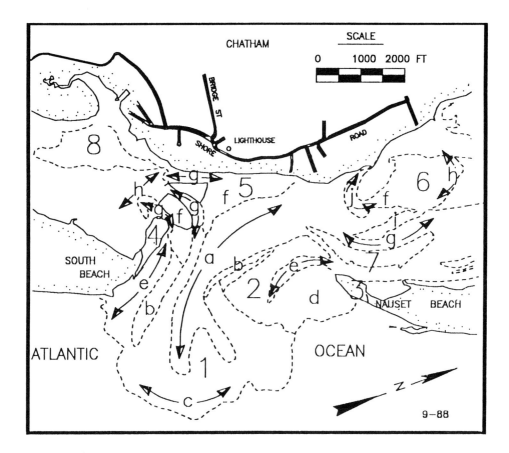

Figure 7. Shoal development of the new inlet (plotted on September, 1988 aerial photograhs). The arrows indicate the inferred sediment transport pathways associated with the shoals. 1 = Ebb Tidal Delta; 2 = Swash Platform with: a = ebb channel; b = channel marginal bars; c = terminal lobe; d = swash bars; e = flood channels; 3 = North Spit; 4 = South Spit; 5 = South Flood-Tidal Delta; 6 = North Flood-Tidal Delta with: f = flood ramp; g = flood channel; h = ebb shield; i = ebb spit; j = spillover lobe; 7 = Remnant Linear Shoal; 8 = Remnant Sand Flat.

The areal changes of the ebb-tidal delta(1) indicate a persistent growth of the shoal into the ocean with a southward shift corresponding to the southward shift in the main channel. The channel-margin swash platform (2) increased rapidly between May and September 1987, and the rate of increase slowed between September, 1987 and September, 1989. The platform area again exhibited a rapid increase to a maximum in May, 1989 (Fig. 8a).

Despite of the periodic changes in its orientation through time, the areal size of the north spit showed little fluctuation except for September, 1987 when

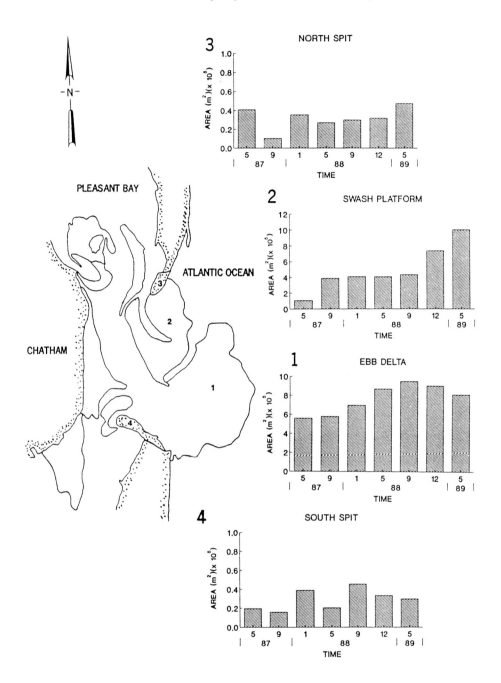

Figure 8a. Post-breach shoal and spit area changes on oceanside of the new inlet (note area scale

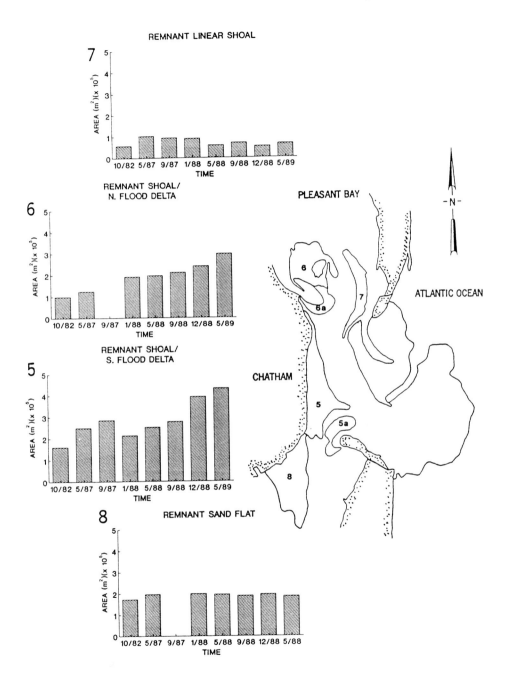

Figure 8b. Post-breach shoal area changes on the landward side of the new inlet.

Table 2. Area change in New Inlet shoals and spits.

DATE:	10/82	5/87	9/87	1/88	5/88	9/88	12/88	5/89	
SHOAL/SPIT SHOAL # TYPE			AREA ($\times 10^5$ m^2)						
1	-	5.60	5.80	7.00	8.74	9.54	9.09	8.24	EBB DELTA
2	-	1.07	3.87	4.08	4.07	4.33	7.41	10.03	SWASH PLATFORM
3	-	0.41	0.10	0.35	0.27	0.30	0.31	0.47	NORTH SPIT
4	-	0.20	0.16	0.39	0.20	0.46	0.33	0.30	SOUTH SPIT
5	1.60	2.47	2.85	2.06	2.90	2.04	1.93	2.57	REMNANT
5a	-	0.03	-	0.07	0.62	0.73	1.99	1.75	SHOALS/S.
Total (5)	1.60	2.50	2.85	2.13	2.52	2.77	3.92	4.32	FLOOD DELTA
6	0.99	1.07	N/A	1.74	1.69	1.64	1.85	2.76	REMNANT
6a	-	0.18	-	0.18	0.28	0.46	0.55	0.27	SHOAL/N.
Total (6)	0.99	1.25	-	1.92	1.97	2.10	2.40	3.03	FLOOD DELTA
7	0.58	1.03	0.94	0.91	0.59	0.70	0.52	0.65	REMNANT LINEAR SHOAL
8	1.74	1.97	N/A	1.99	1.96	1.89	1.96	1.87	REMNANT SAND FLAT

the area of the north spit was reduced to about half of its average size. Between May, 1987 and May, 1989, there was a small net decrease in area of the north spits. The area of the south spit fluctuated as a result of the cyclical growth and periodic separation of the distal end of the spit. However, a net gain of 0.1×10^5 m^2 was measured on the south spit during the length of the study.

The evolution of the flood-tidal deltas is dated back to shoals that existed in the lagoon prior to the breach. As the inlet evolved, these shoals took on more characteristics of inlet-related configurations. In 1982, a large sand flat (Fig. 4a) approximately of 1.6×10^5 m^2 in area existed off the Chatham lighthouse on the mainland shore. After the breach, it has been modified by waves and tidal currents and became incorporated into the south flood-tidal shoal complex (5 & 5a, Fig. 8b). This shoal complex continued to grow to a maximum size by May, 1989. The interaction of this shoal with the growth of the south spit caused the area of the shoal to fluctuate. The shoal complex has essentially prevented effective water exchange between the northern and southern parts of the lagoon. A net gain of approximately 2.71×10^5 m^2 was measured from October, 1982 to May, 1989, as this south flood-tidal shoal continued to evolve and migrated southward.

A large shoal (6) having the area of 1.0×10^5 m^2 also existed in mid-lagoon in 1982, which caused the main channel to bifurcate on its way to Pleasant Bay (Fig. 4a). After the breach, this shoal has been modified to become the north flood-tidal delta complex (6 & 6a, Fig. 8b). There has been steady growth in the area of this shoal with time (Fig 8b.). No area was measured from the September, 1987 photographs since the entire shoal was not visible on the lower altitude photo set. A net gain of approximately 1.99×105 m2 was measured from October, 1982 to May, 1989. The growth of an ebb spillover lobe (6a) extending into the original channel may become a hazard for navigation as this shoal complex continues to develop.

The linear sand shoal (7) gained approximately 0.5×10^5 m^2 of area after the breach and has increased slightly by 0.07×10^5 m^2 over the twenty-four month period since May, 1987 (Fig 8b). The other remnant sand flat area (8) increased slightly in area immediately after the breach and has remained essentially constant during the study period. The transport of sediment south-ward along the mainland shoreline, and the slow movement of the south flood-tidal delta into the southern part of the lagoon, indicate that more sand will probably reach this shoal in the near future.

Southward Migration of the Main Inlet Channel

Aerial photographs show that a main channel between the north and south spit (not well-defined on May, 1987 photographs, Fig. 4b) has formed, which bends seaward from shore-subparallel to shore-perpendicular through the new inlet mouth. The changes of the channel orientation suggest the capture of the tidal prism by the newly formed inlet. The water exchange between the estuary and ocean is now more efficient through the new inlet than through South Channel, due to less friction (shorter route).

South of the main channel bend is the area of an extensive subtidal shoal complex. These shoals formed a platform on which the detached remnants of the south spit underwent transformation to become flood tidal-delta-like shoals. Interspersed between these intertidal islands that extended across the estuary are small flood-dominated secondary channels connecting the two parts of the estuary (Fig. 4e-h). These channels are only effective during the late stage of the flood tide and early stage of the ebb tide when the water elevation is high.

The existence of the broad and shallow inter- to subtidal channel-margin swash platform on the updrift side of the main channel in the throat area resulted in an asymmetric configuration shown by the cross-sectional profile (marked by A-A' in Fig. 4e) of the inlet mouth (Fig. 9). The position of the southernmost edge of this platform showed southward advancement from May, 1987 to May, 1989 (represented by the position of the northern bank of the main channel throat, Fig. 6). Because of the net northward retreat of the north spit, the length of this platform actually increased.

As the edge of the channel-margin swash platform progressively advanced southward, the position of the mid-point of the main channel in the inlet throat also shifted southward. This southward movement has an average migration rate of 12 m per month between May, 1987, and September, 1988 (Fig. 6), and a slower rate of 9 m per month from September, 1988 to May, 1989.

From May, 1987, to May, 1988, the width of the channel in the throat section increased from approximately 110 m to 192 m, and remained about the same thereafter. Since the width of the channel is proportional to the size of the channel (Bruun et al., 1978), the asymptotic trend of the channel width increase may suggest that the expansion of the channel in the inlet throat has stopped between May and September, 1988. In other words, the main channel is no longer in the scouring stage, characteristic of the initial stage of inlet development (Bruun et al., 1978). However, the channel, while maintaining a constant width through the throat area, is still migrating southward with the ebb-tidal delta, causing the south spit to retreat southward.

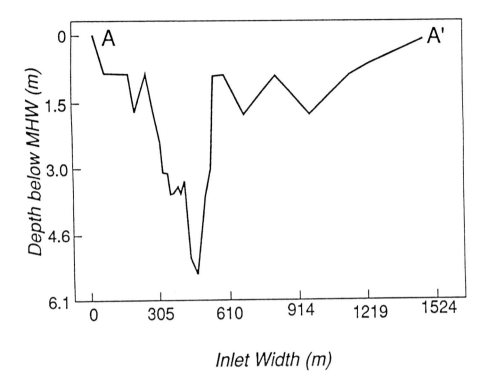

Figure 9. Cross-sectional profile across the inlet mouth (see Fig. 4e for position).

Sediment Patterns along the Shoals South of the Main Channel Bend

Eight surficial sediment samples were collected in a transect across the south spit and flood tidal shoal/channel complex (Fig. 10, Table 3). The flood-tidal delta sediments all group in the well sorted and medium mean grain size category. Within this grouping, the sub-environments of the overwash (#1) and tidal flat samples (#5, 7) are distinguished by slightly better sorting and slightly coarser means than the flood channel samples (#4, 6, and 8). The overwash throat on the spit was probably influenced by waves and tidal currents from the mid to high portion of the tidal cycle. The two tidal flat areas

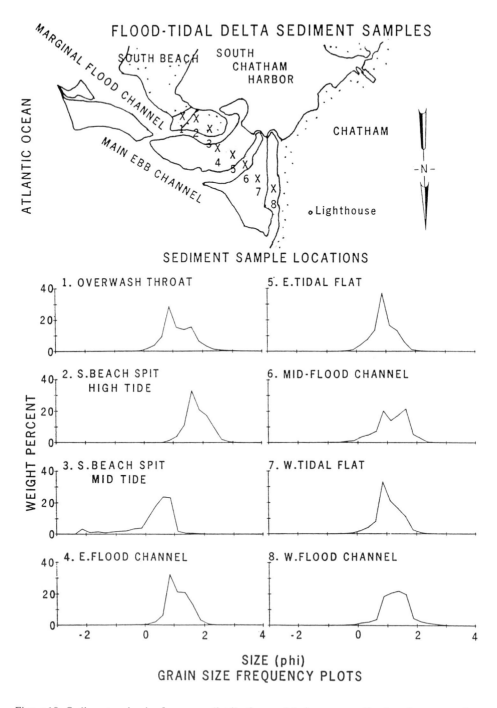

Figure 10. Sediment grain-size frequency distributions and their corresponding locations across the shoals between the south spit and the inner shoreline at low water in September, 1988.

Table 3. Sediment sample location and grain size statistics

| Location | Moment Mean | | Moment Sorting | Moment Skewness | Moment Kurtosis |
	(Φ)	mm	(Φ)	(Φ)	(Φ)
Overwash	1.13	0.46	0.36	0.00	3.47
High Tide on Spit	1.80	0.29	0.37	0.12	3.16
Mid Tide on Spit	0.29	0.82	0.76	-1.60	5.59
East Flood Channel	1.18	0.44	0.51	0.33	3.78
East Tidal Flat	0.92	0.53	0.40	-0.17	3.67
Mid Flood Channel	1.15	0.45	0.53	-0.59	3.42
West Tidal Flat	1.07	0.49	0.41	-0.01	4.04
West Flood Channel	1.27	0.42	0.49	-0.01	4.9

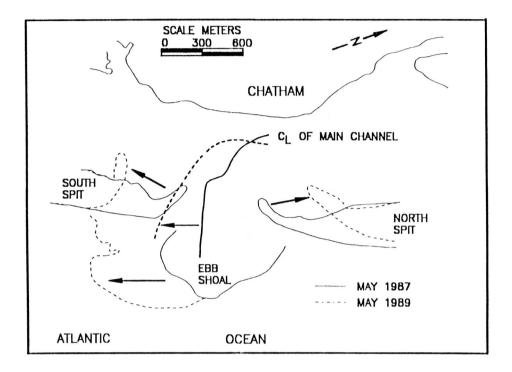

Figure 11 Net changes in north and south spit shorelines, ebb-tidal delta, and thalweg (C_L) of main channel from May, 1987 to May, 1989.

are under the influence of tidal currents around high tides and are exposed during low tides. Each channel sample exhibited slightly bimodal distributions, perhaps suggesting the unequal influence of bi-directional tidal currents.

Processes

Based on the evolution of the inlet and shoal morphology and sediment distribution and channel bathymetry, processes driving this system can be inferred. Morphological changes in the development of the new inlet from May, 1987, to May, 1989, are summarized in Figure 11, which highlights spatial changes and direction of movements observed in the two barrier spits, ebb-tidal delta, and main channel.

The tip of the north spit moved approximately 700 m northward and slightly landward while the south spit moved about some 580 m southward and appreciably landward due to the shore-normal re-orientation. The ebb-tidal delta has grown seaward and moved to the south while a large channel-margin swash platform on the north and a marginal linear bar on the south of the main channel throat have developed within the inlet mouth. The main channel thalweg also shifted position some 400 m to the south. The south spit/south flood-tidal delta complex has virtually filled in the N-S oriented channel that existed in May, 1987, as the delta moves farther southward into the estuary. The remnant shoal/north flood-tidal delta complex has developed and seems to cause shoaling in the adjacent navigation channel.

The Influence of Waves and Tidal Currents

Seasonal wave climate and longshore drift patterns may play a role in the fluctuations of both the north and south spit development. The southwesterly waves and northward drift in summer caused the north spit to retreat farther to the north, whereas in winter the predominantly stronger northeasterly waves and southward drift erode the south spit into the lagoon. Wave refraction around the growing ebb-tidal delta is also likely to play a role in forming a localized northward drift aiding in the south spit elongation into the lagoon.

The apparent downdrift migration of the main channel and inlet throat, the edge of the updrift channel-margin platform, and the downdrift retreat of the south spit (north end of South Beach) all indicate sediment accumulation on the updrift side of the main channel thalweg and concurrent sediment removal on the downdrift side. This depositional pattern suggests trapping of littoral sand within the area of the new inlet mouth and resulting insufficient sand by-passing (Oertel, 1988).

In addition, sediments from the littoral drift are partially diverted seaward and landward by tidal flows through the main channel to form the flood and ebb-tidal deltas. The influx of littoral sediments into the lagoon is probably aided by waves and is evident from the accretion on the inner shore south of the main channel bend (Giese et al., 1989), from the cyclical elongation of the south spit, and from the extensive growth on the north and south flood-tidal delta complexes. The depositional processes along the shoals between the south spit and the mainland shore south of the main channel bend are influenced both by ocean waves and tidal currents. Sediment grain size characteristics along these shoals indicate strong wave influence on the landward-accreting south spit, and moderate wave influence on the tidal flats. Among the secondary channels that connect the main channel and South Chatham Harbor, the landward- and seaward-most secondary channels are probably dominated by flood-tidal currents. The mid-secondary channel is probably more affected by ebb tidal currents that flow from South Chatham Harbor into the main channel of the new inlet. Despite of the influx of ocean sediments and the prevalent depositional patterns inside the estuary, the small erosional segment on the inner shore (Giese et al., 1989), however, is a localized phenomenon, which is caused by the proximity of the main channel to the shore as shown by the cross-sectional profile of the channel (marked B-B' in Fig. 4e) in Figure 12.

Morphodynamics of the Main Channel of the New Inlet

Although a hydrodynamic equilibrium is attained at the inlet throat, the continuing southward migration of the main channel and the ebb-tidal delta suggests that the position of the inlet has not yet reached a state of morphodynamic equilibrium in the entire inlet/barrier-bay system. The correspondence between the accretional trend of the edge of the channel-margin platform and the erosional trend of the south spit (Fig. 6) indicates that these two features are both related to the southward-advancing main channel throat that separates the two. Aubrey and Speer (1984) suggested that a bending inlet channel configuration can be responsible for lateral inlet migration, creating a steep outer (southern, in the case of this study) channel bank, and an accreting point bar on the inner (northern) bank. This mechanism is likely operational in the new inlet of Chatham due to its morphological similarities to a bending channel. However, because of the effects of littoral drift and shoaling waves, the point bar equivalent in the Chatham case is the broad updrift channel-margin swash platform. Subsequently, this channel bending mechanism is the dominant factor in causing the southward migration of the main channel

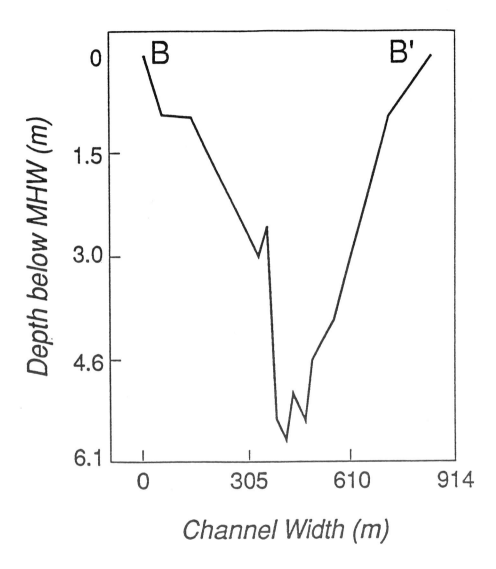

Figure 12. Cross-sectional profile across the main channel between mainland beach and north spit (see Fig. 4e for position).

throat. In other words, the southward movement of the channel throat is largely controlled and sustained by the configuration of the channel thalweg. This process can probably continue without interference from the littoral drift system as long as the channel geometry remains unchanged.

The northward trend for the tip of the north spit suggests that the movement of the north spit is unrelated to the movement of the main channel throat. Since the south end of North Beach is on average about 1 km from the channel throat section and is separated by a channel-margin platform, it is conceivable that processes causing southward movement of the throat and the lengthening of the channel-margin swash platform are not directly affecting the north spit. It is likely that the net shoreline retreat on the south end of North Beach is caused by the seaward re-orientation and displacement of the nearshore bar, which is a littoral sediment conduit, to join the distal end of the ebb-tidal delta off the south end of North Beach (Fig. 4). Consequently, the south end of North Beach is experiencing a net sediment deficit, which is occasionally offset by the change in the position of the nearshore bar and the nearshore wave field. The northward retreat of the north spit will gradually stop when the inlet throat is far enough from the spit so that the effect of the ebb-tidal delta on the nearshore bar system no longer affects the south end of North Beach.

Sediment Transport Pathways

Based on the developments ebb- and flood-tidal shoals, a hypothetical sediment transport pathway map has been constructed (Fig. 13). Sand enters the inlet from the north in the longshore transport system along the shoreline and nearshore bar of Nauset Beach. Some sand is transported into the inlet across the channel-margin swash platform via swash bar migration, wave transport, and in the associated marginal channels around the north spit. Some sand is moved into the main channel where it is transported both landward and seaward. Localized northward transport adjacent to the South Beach shore-line moves sand into the lagoon in the south marginal flood channel along the south spit. Some material may be by-passed around the ebb-tidal delta and transported along South Beach towards the shoals and spit associated with South Channel to the south.

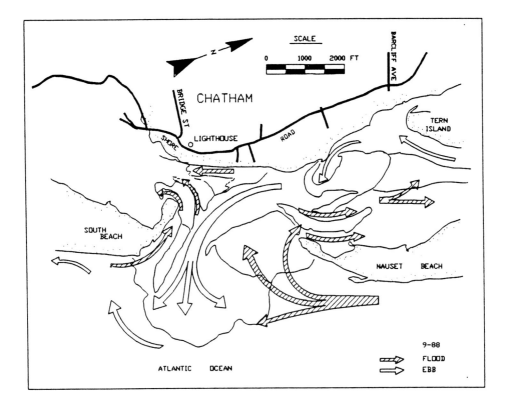

Figure 13. Hypothetical sediment transport pathways at the new inlet based on morphology from September, 1988 shoal configuration.

Inside the inlet mouth, the sediment entering from the north side of the main channel flows northward along the linear shoal and across the north flood-tidal delta towards Pleasant Bay. The sand entering from the main channel and south marginal flood channel built the south flood-tidal delta, which continues to grow and move southward into South Chatham Harbor. The ebb shields developing on the southern end of the three secondary channels on the south flood-tidal delta complex will progressively block northward ebb flow during the latter portion of the ebb cycle from flowing back into the main channel.

Conclusions

A new tidal inlet has been created by breaching of a barrier spit during a storm. The presence of this inlet interrupts the old barrier/lagoon system. As result, adjacent shorelines of the newly formed inlet have responded to the new hydrodynamic conditions. Since its formation, the mouth of the new inlet has been widening through time. This widening is contributed by the concurrent retreats of the two spits that flank the new inlet. Along with the progressive retreat, the north spit displayed fluctuations in its length and orientation. The south spit on the other hand, although maintaining a shore-normal orientation, has shown cyclical elongation followed by periodic breaching and terminal detachment.

As the new inlet evolved, ebb- and flood-tidal shoal complexes have formed. Six shoal areas have been identified. Seaward of the new inlet mouth, there is a well developed ebb-tidal delta. Within the inlet mouth, a large channel-margin swash platform is largely located on the updrift side of the channel throat of the new inlet. Both features have grown in size as the inlet evolved. As the main channel migrated to the south, the ebb-tidal delta has grown both seaward and southward.

Four other shoal features have been identified to be flood-tidal delta complexes of the new inlet. A linear shoal and northern flood-tidal delta located on the northern side of the new inlet evolved from remnant features now modified by the new hydrodynamic conditions since the breach. The south flood-tidal delta is growing in area and also moving into the southern portion of Chatham Harbor. The south flood-tidal delta effectively prevented the water exchange between the northern and southern parts of the estuary. The growth of shoals landward of the inlet mouth and the cyclical elongation of the south spit suggest the influx of ocean sediments entering the lagoon through the new inlet.

As a result of the shorter exchange route and the capture of the most of the tidal prism, the shore-parallel main channel thalweg turned seaward at a right angle to form a bend through the new inlet mouth. This new channel configuration is likely responsible for the southward migration of the channel throat and partially for the formation of the shallow channel-margin swash platform on the updrift side of the channel throat. The continuing southward migration of the new inlet channel and associated shoals suggests the morphodynamics of the new inlet have not reached equilibrium with the entire barrier/lagoon system. Furthermore, the northward movement of the north spit indicated the north spit is not influenced by the same processes that caused the rest of the new inlet system to move southward.

Morphological changes following the formation of the new inlet suggest that Chatham Harbor has developed into two separate parts. The northern part of the lagoon is actively interacting with the littoral system and increasing in length as the channel throat advances southward at the expense of the southern part of the estuary. The mouth of the new inlet effectively traps sediments from the littoral system, and less sediment is likely to by-pass the new inlet in the immediate future.

Acknowledgments

This study was partially funded by the Town of Chatham, the Commonwealth of Massachusetts through the Department of Environmental Management, the Coastal Engineering Research Center of the U.S. Army Corps of Engineers, the Coastal Research Center of Woods Hole Oceanographic Institution (WHOI), and NOAA National Sea Grant No. NA86-AA-D-SG-90, WHOI Sea Grant Project No. R/O-6. The U.S. Government is authorized to produce and distribute reprints for governmental purposes notwithstanding any copyright notation that may appear hereon. Permission was given by the Office of Chief of Engineers to publish this paper. We thank Dr. Gary Zarillo and Dr. John Boothroyd for reviewing an early version of this manuscript. Woods Hole Oceanographic Institution Contribution No. 8316.

REFERENCES

Aubrey, D.G. and P.E. Speer, 1984. Updrift migration of tidal inlets. *J. Geol.*, v. 92, p. 531-546.

Aubrey, D.G., D.C. Twichell and S.L. Pfirman, 1982. Holocene sedimentation in the shallow nearshore zone off Nauset Inlet, Cape Code, Massachusetts. *Mar. Geol.*, v. 47, p. 243-259.

Boothroyd, J.C., 1985. Tidal inlets and tidal deltas. In: *Coastal Sedimentary Environments*, Davis, R.A. Jr. (ed.), Springer-Verlag, New York, N.Y., p. 445-532.

Bruun, P., A.J. Mehta, and I.G. Johnsson, 1978. *Stability of Tidal Inlets.* Elsevier Scientific Publ. Co., New York, 510 pp.

Cornillon, P., 1979. Computer simulation of shoreline recession rates, outer Cape Cod. In: *Environmental Geological Guide to Cape Cod National Seashore*, Leatherman, S.P. (ed.), Field Trip Guide Book for the Eastern Section of SEPM, p. 41-54.

Ebert, J.R. and C.R. Weidman, 1989. Inlet-spits and island/shoal calving: a cyclic process in the development of a flood-tidal delta, Cape Cod, Massachusetts. Abstract, *NE Geol. Soc. Amer.*

FitzGerald, D.M., 1988. Shoreline erosional-depositional processes associated with tidal inlets. In: *Hydrodynamics and Sediment Dynamics of Tidal Inlets,* Aubrey, D.G. and L. Weishar (eds.), *Lecture Notes on Coastal and Estuarine Studies,* v. 29, Springer-Verlag, New York, p. 186-225.

Friedrichs, C.T., D.G. Aubrey, G.S. Giese and P.E. Speer. Hydrodynamical modeling of a multiple-inlet estuary/barrier system: Insight into tidal inlet formation and stability. In: D.G. Aubrey and G.S. Giese (eds.), *Formation and Evolution of Multiple Inlet Systems, Coastal and Estuarine Studies,* (this volume).

Giese, G.S., 1988. Cyclical behavior of the tidal inlet at Nauset Beach, Chatham, Massachusetts. In: *Hydrodynamics and Sediment Dynamics of Tidal Inlets,* Aubrey, D.G. and L. Weishar (eds.), *Lecture Notes on Coastal and Estuarine Studies,* v. 29, Springer-Verlag, New York, p. 269-283.

Giese, G.S., D.G. Aubrey, and J.T. Liu, 1989. Development, Characteristics and Effects of the New Chatham Harbor Inlet. Technical Report, Woods Hole Oceanographic Institution, WHOI-89-19, 32 pp.

Hayes, M.O., 1980. General morphology and sediment patterns in tidal inlets. *Sedimentary Geology,* v. 26, p. 139-156.

Leatherman, S.P., 1979. Overwash processes on Nauset Spit. In: *Environmental Geological Guide to Cape Cod National Seashore*, Leatherman, S.P. (ed.), Field Trip Guide Book for the Eastern Section of SEPM, p. 171-192.

Leatherman, S.P., 1988. Cape Cod Field Trip. Coastal Publ. Series, U. of Maryland, 132 pp.

McClennen, C.E., 1979. Nauset Spit: model of cyclical breaching and spit regeneration during coastal retreat. In: *Environmental Geological Guide to Cape Cod National Seashore*, Leatherman, S.P. (ed.), Field Trip Guide Book for the Eastern Section of SEPM, p. 109-118.

Oertel, G.F., 1988. Processes of sediment exchange between tidal inlets, ebb deltas, and barriers islands. In: *Hydrodynamics and Sediment Dynamics of Tidal Inlets*, Aubrey, D.G. and L. Weishar (eds.), *Lecture Notes on Coastal and Estuarine Studies,* v. 29, Springer-Verlag, New York, p. 297-318.

Weishar, L., D. Stauble, and K. Gingerich, 1989. A Study of the Effects of the New Breach at Chatham, Massachusetts. Reconnaissance Report, Coastal Eng. Res. Center, Army Corps of Engineers, 164 pp.

Wright, A.E. and B.M. Brenninkmeyer, 1979. Sedimentation patterns at Nauset Inlet, Cape Cod, Massachusetts. In: *Environmental Geological Guide to Cape Cod National Seashore*, Leatherman, S.P. (ed.), Field Trip Guide Book for the Eastern Section of SEPM, p. 119-140.

3

Hydrodynamical Modeling of a Multiple-Inlet Estuary/Barrier System: Insight into Tidal Inlet Formation and Stability

Carl T. Friedrichs, David G. Aubrey, Graham S. Giese and Paul E. Speer

Abstract

Two specific questions are addressed concerning the role of tidal hydrody-namics in determining the long-term morphologic evolution of the Nauset Beach-Monomoy Island barrier system and the Chatham Harbor-Pleasant Bay tidal estuary, Massachusetts: (1) why do the barrier and estuary exhibit a long-term (~150 yr) cycle of new inlet formation, and (2) once a new inlet forms, why is the resulting multiple inlet system unstable? To address these questions, a branched 1-d numerical model is used to recreate the basic flow patterns in the tidal estuary at ten-year intervals during the last half century and also to recreate flow conditions shortly before and shortly after the formation of the new inlet. Results suggest that an inlet will form through Nauset Beach once southerly elongation of the barrier has led to a critical head across the barrier at high tide. If this critical head (enhanced by storm surge and wave set-up) exists at high tide during consecutive tidal cycles, flood currents can deepen the overwash channel sufficiently to enable the stronger ebb currents to complete the formation process. Once a new inlet has formed, the surface gradient and tidal discharge are drastically reduced along the pre-existing channel to the south of the inlet. This reduction eliminates the tidal

Formation and Evolution of Multiple Tidal Inlets
Coastal and Estuarine Studies, Volume 44, Pages 95-112

scouring action needed to keep the channel open. Rapid shoaling within the channel to the south of the new inlet completes the hydrodynamic decoupling of the northern and southern sections of the estuary.

Introduction

Tidal inlets between barrier spits and/or barrier islands change continually. Unless restricted by engineering structures or by naturally occurring obstacles such as resistant subsurface lithologies (FitzGerald and FitzGerald, 1977), they commonly migrate alongshore, frequently - but not always - in a downdrift direction (Aubrey and Speer, 1984). During severe storms accompanied by unusually high sea levels and waves attacking the outer barrier, new inlets may form and pre-existing inlets may close. Under most circumstances the general form and structure of the barrier through which the inlets pass remain intact despite such changes in the inlets themselves. However, an entirely different situation can be found in cases involving inlets situated at the downdrift end of barrier systems. Here, the barrier spit or island itself may

Figure 1. Aerial photograph of New Inlet on 3 January 1987 one day after its formation during a severe easterly storm with perigean spring high tides. The distance across the breach at this time was on the order of 100 meters. By 1991, the inlet width exceeded 2 km.

become unstable, and drastic changes in barrier form may be triggered by changes in associated tidal inlets. A particularly clear example of the role of such "terminal" inlets is provided by the relationship between New Inlet in Chatham, Massachusetts (Fig. 1) and the patterns of change exhibited by the Nauset Beach-Monomoy Island barrier system.

A striking feature of this barrier system is its long period (150 yr) cycle of change (Fig. 2). In its simplest form, the system consists of two unbroken barriers, Nauset Beach and Monomoy Island, and all tidal flow between the ocean and the Chatham Harbor-Pleasant Bay estuary passes through a single tidal inlet (South Inlet) located immediately south of Nauset Beach (Fig. 2, c.1920, c.1940). However, as littoral drifting causes Nauset Beach to elongate to the south, Monomoy Island separates from Morris Island and a second inlet (West Inlet) is created (Fig. 2, c.1960, c.1980). Later, after continued southward growth, Nauset Beach is breached, forming a third inlet (New Inlet—Fig. 2, c.1990; Fig. 3). Following this event, the separated south end of Nauset Beach migrates onto shore, infilling the old tidal channel (Middle Channel) in the process. Eventually the migrating sand mass reconnects Monomoy Island to Morris Island and recreates the initial configuration.

The cyclical behavior of this barrier system has been discussed frequently (e.g., Mitchell, 1874; U.S. Army Corps of Engineers, 1968; Oldale et al., 1971; Goldsmith, 1972; McClennen, 1979; Aubrey, 1986; Leatherman and

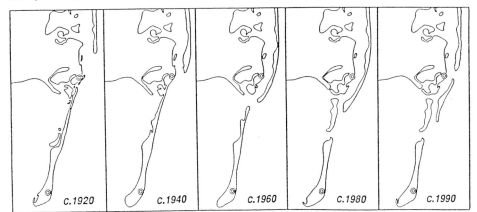

Figure 2. Historical Changes in the Nauset Beach-Monomoy Island barrier system during the most recent barrier growth cycle (after Giese, 1988). The sites of the Chatham and Monomoy Lighthouses are indicated.

Zaremba, 1986; Giese, 1978, 1988; Liu et al., this volume). In this paper we
address two specific questions regarding the role of tidal hydrodynamics in
the evolution of the Nauset Beach-Monomoy Island barrier system and the
Chatham Harbor-Pleasant Bay tidal estuary: (1) why do the barrier system

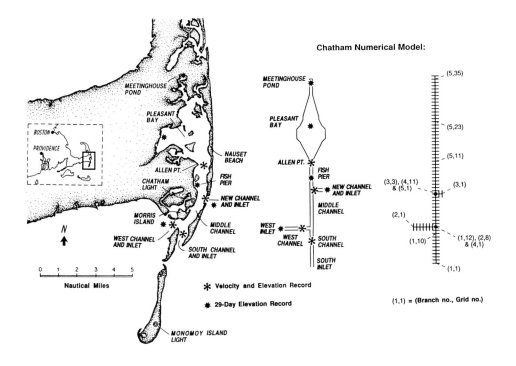

Figure 3. The Nauset Beach-Monomoy Island barrier system and Chatham Harbor-Pleasant Bay
estuary under 1988 conditions with the locations of pressure sensors and current meters which recorded
the data presented in Table 1. Also shown is a schematic layout of the one-dimensional, branched
numerical model with corresponding sampling locations and a line-drawing of the model indicating the
branch and grid numbering scheme.

and tidal estuary exhibit this long-term cycle of new inlet formation, and (2) once a new inlet forms, why does Middle Channel become unstable, resulting in the decoupling of the northern and southern sections of the estuary?

The following section describes the numerical modeling techniques used in this study. The next section describes the tidal data used to force the model and compares observed and modeled tidal elevations and velocities from throughout the system. In the final two sections we use the model to examine specifically the cyclical behavior of barrier breaching and the stability of the multiple-inlet system following barrier breaching.

Model Formulation

For this study we adapted an existing one-dimensional nonlinear tidal propagation model for use in a multiple inlet system. Earlier versions have been applied successfully to single inlet, single channel applications (Speer and Aubrey, 1985; Friedrichs and Aubrey, 1988) and to branched channel applications (Friedrichs and Aubrey, 1989; Giese et al., 1989b). The model is based on the cross-sectionally integrated equations of motion:

$$b\frac{\partial \zeta}{\partial t} + \frac{\partial Q}{\partial x} = 0, \qquad \frac{\partial Q}{\partial t} + \frac{\partial}{\partial x}\frac{Q^2}{A} = -gA\frac{\partial \zeta}{\partial x} - c_d \frac{Q}{A}\left|\frac{Q}{A}\right|P,$$

where b = channel width, ζ = water surface elevation, Q = volume transport, A = channel cross-sectional area, g = acceleration of gravity, c_d = drag coefficient, and P = wetted channel perimeter. The above representation of estuarine physics assumes estuary length >> width, width >> depth, a well-mixed water-column, and negligible fresh-water inflow.

To solve the above equations, continuous derivatives were replaced by centered differences in space and forward differences in time:

$$b_j \frac{\zeta_j^{n+1} - \zeta_j^n}{\Delta t} + \frac{Q_{j+1}^n - Q_{j-1}^n}{2\Delta x} = 0, \qquad (1)$$

$$\frac{Q_j^{n+1} - Q_j^n}{\Delta t} + \frac{1}{2\Delta x} \left(\frac{\left(Q_{j+1}^n\right)^2}{A_{j+1}^n} - \frac{\left(Q_{j-1}^n\right)^2}{A_{j-1}^n} \right) =$$

$$- gA_j^{n+1} \frac{\zeta_{j+1}^{n+1} - \zeta_{j-1}^{n+1}}{2\Delta x} - c_d \frac{Q_j^n}{A_j^n} \left| \frac{Q_j^n}{A_j^n} \right| P_j^n . \tag{2}$$

where Δt and Δx indicate the time and space step sizes, subscript j indicates grid number, and superscript n indicates time step number. At each new time step, (1) was solved first for sea surface elevation, ζ_j^{n+1}. Then (2) was solved for transport, Q_j^{n+1}. A_j^n and P_j^n were found as simple functions of sea surface elevation. The form of the advective term used in this numerical scheme can cause numerical instability under certain circumstances. In this application, however, dissipation through the friction term overcame any tendency towards advective instability.

The model is composed of one-dimensional branches which connect at nodes (Fig. 3). The model was forced by prescribed, periodic time series of elevation at each inlet, and a no flow condition was applied at inland boundaries. Two matching conditions were applied at each node: continuity of transport ($Q_1 + Q_2 + Q_3 = 0$) and continuity of elevation ($\zeta_1 = \zeta_2 = \zeta_3$). At the inlets, inland boundaries and nodes, centered differences in space were replaced by either forward or backward differences (i.e., $j+1$ or $j-1$ was replaced by j, and $2\Delta x$ was replaced by Δx), as appropriate. Throughout this study we used $\Delta t = 15$ seconds, $\Delta x = 250$ meters, and $c_d = 0.02$, scales previously found to be appropriate for shallow, frictionally-dominated tidal embayments (Speer and Aubrey, 1985; Friedrichs and Madsen, 1992).

The numerical modeling performed in this study was diagnostic in nature rather than predictive. Our aim was to determine which aspects of the barrier and estuary geometry control the fundamental hydrodynamic patterns and therefore most strongly influence patterns of morphologic evolution. Our aim was not to reproduce exactly observed time series of tidal elevation and/or velocity. Therefore, to isolate the geometric features of interest, we chose to model the tidal estuary simply, using constant depth, rectangular channels.

We chose a uniform depth of 2.5 meters (relative to mean sea level at the inlets), which is approximately the average depth of the Chatham Harbor-Pleasant Bay estuary. We also chose a constant channel width of 300 meters throughout the embayment, except in the immediate vicinity of Pleasant Bay, where the model width increased to 2000 meters (Fig. 3). The model was not "tuned" in any way.

Comparison with Field Data

Field data were collected as part of a larger observational study which investigated the overall morphological response of the barrier system to the formation of New Inlet (Giese et al., 1989a; Liu et al., this volume), and data acquisition methods are described in greater detail elsewhere. Synoptic observations of tidal elevation in April and May 1988 were provided by five temperature and depth recorders (TDRs) deployed at locations indicated in Figure 3 and Table 1. Additional elevation data were provided by a TDR deployment in Pleasant Bay in September 1988. During April and May, 1988, current meters, which also recorded tidal elevation, were deployed at four additional locations (Fig. 3, Table 1). Because of equipment limitations, the current measurements could not be synoptic (Giese et al., 1989a). Elevation set-up could not be estimated from field data because of inadequate surveying between instruments.

To approximate the mean conditions observed in the field, model elevations were forced at both South Inlet and New Inlet using the M_2 and M_4 tidal components observed offshore of New Inlet in April 1988 (Table 1). West Inlet, which faces Nantucket Sound, was forced with the M_2 and M_4 components observed offshore of West Inlet (Table 1). A visual comparison of field observations and numerical model results (Fig. 4) indicates that the simple numerical model captured the fundamental hydrodynamic behavior of the tidal estuary. For 1988 conditions, both model results and observations indicate a progressive wave relation between elevation and velocity at Allen Point, a standing wave relation in West Channel, and an intermediate relation in New Channel. The modeled and observed elevation-velocity relation in South Channel (not shown) are also consistent and similar to the intermediate relation in New Channel. A quantitative comparison of numerical results and observations for 1988 conditions is presented in Table 1.

The observed and modeled M_2 elevations agree well throughout almost all of the system. The elevation observations collected by pressure sensors de-

Table 1.

Elevations: observed and modeled (in parentheses)

gauge or branch, grid	starting date	duration in hours	M_2 amp. in cm	M_2 lag in deg	M_4/M_2 amp. ratio	$2M_2$-M_4 phase in deg	mean elev. in cm
West Inlet (2,1)	5 Apr. '88	697	56 (56)	8 (8)	0.086 (0.086)	252 (252)	n.a. (0)
West Channel (2,6)	6 Apr. '88	73	73 (74)	n.a. (8)	0.025 (0.030)	349 (336)	n.a. (4)
South Channel (1,10)	4 Apr. '88	73	79 (84)	n.a. (7)	0.004 (0.034)	186 (14)	n.a. (4)
Ebb-tide delta (3,1)	5 Apr. '88	697	105 (105)	0 (0)	0.025 (0.025)	285 (285)	n.a. (0)
New Channel (3,3)	21 Apr. '88	73	80 (83)	31 (8)	0.033 (0.045)	29 (19)	n.a. (6)
Fish Pier (5,4)	5 Apr. '88	697	66 (68)	35 (21)	0.052 (0.066)	75 (37)	n.a. (10)
Allen Point (5,10)	21 Apr. '88	232	66 (54)	49 (51)	0.074 (0.108)	74 (46)	n.a. (13)
Pleasant Bay (5,23)	22 Sep. '88	697	54 (54)	69 (60)	0.159 (0.133)	64 (50)	n.a. (13)
Meeting House Pond (5,35)	5 Apr. '88	697	59 (54)	73 (60)	0.219 (0.134)	73 (53)	n.a. (13)

Along-channel velocities: observed and modeled (in parentheses)

gauge	starting date	duration in hours	M_2 amp. in cm/s	elev.-vel. phase in deg	M_4/M_2 amp. ratio	$2M_2$-M_4 phase in deg	mean vel. in cm/s
West Channel (2,6)	6 Apr. '88	73	82 (45)	4 (0)	0.152 (0.188)	76 (41)	26 (7)
South Channel (1,10)	4 Apr. '88	73	38 (35)	27 (21)	0.112 (0.102)	325 (353)	- 3 (1)
New Channel (3,3)	21 Apr. '88	73	106 (78)	38 (37)	0.098 (0.118)	267 (340)	- 8 (- 3)
Allen Point (5,10)	21 Apr. '88	73	59 (57)	63 (81)	0.216 (0.223)	357 (349)	-16 (- 8)

ployed with the current meters may not represent equilibrium conditions because their record lengths are relatively short. The observed and modeled M_4 elevations and velocities are consistent in terms of both relative phase and amplitude, which indicates the numerical model also captured the fundamental nonlinear tidal processes occurring in the tidal estuary. The amplitudes of observed and modeled M_2 velocities were expected to disagree to some degree for at least three reasons: (1) the short velocity records do not represent equilibrium flow conditions; (2) model results are cross-sectionally averaged velocities whereas observations are point velocities; and (3) the model did not attempt to represent the smaller scale expansions and contractions in cross-sectional geometry. Nevertheless, the comparison of numerical and observational data presented in this section indicates that a relatively simple model captures the fundamental linear and nonlinear hydrodynamics observed at the tidal estuary in 1988.

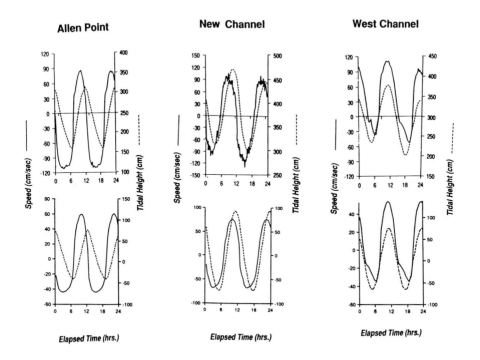

Figure 4. Comparison of observed and modeled tidal elevations and along-channel velocities for three locations within the tidal estuary (see Figure 3). The upper three plots are observed records and the lower three plots are modeling results. Positive velocities are northward at Allen Point and westward at New Channel and West Channel.

Examination of Cyclical Behavior

To investigate the role of tidal hydrodynamics in cyclical barrier breaching, we used the numerical model to recreate the basic flow patterns in the Chatham Harbor-Pleasant Bay estuary at ten-year intervals between 1936 and 1986 (first six configurations in Fig. 5). The historical geometries are based on previous studies of the morphologic evolution of the Nauset Beach-Monomoy Island barrier system (Giese, 1978, 1988). During the first three ten-year periods (Fig. 5), the system had only one inlet, and the corresponding models were forced only by the Atlantic M_2 and M_4 tides (as observed off New Inlet in May 1988 — see Fig. 3). In 1958 a breach formed between Monomoy Island and the mainland, creating a second inlet into the estuary. Thus the next three models (Fig. 5) contain two inlets: West Inlet and South Inlet. For these three models, West Inlet was forced with the Nantucket Sound M_2 and M_4 tides (as observed off West Inlet in May 1988), whereas South Inlet was forced with the Atlantic tide.

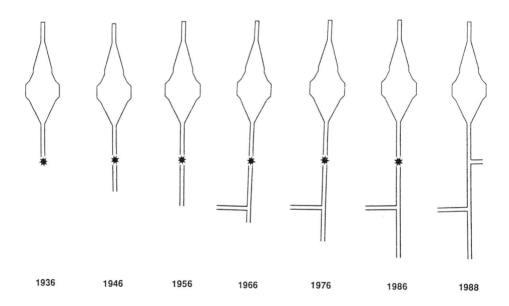

| 1936 | 1946 | 1956 | 1966 | 1976 | 1986 | 1988 |

Figure 5. Schematic layouts of numerical models representing the Chatham Harbor-Pleasant Bay estuary at various points during its evolution, based on historical data summarized in Figure 1. Asterisks indicate the site of tidal elevation and tidal head calculations displayed in Figure 6.

We examined the elevations produced by the six historical models at the site of the 1987 breach under mean tidal conditions (Fig. 6a). Tidal evolution is as might be expected, consisting of monotonically reduced tidal range, increasing phase lag, increasing mean water level set-up, and increasing tidal nonlinearity (larger M_4/M_2 ratio). From the modeled elevations, we calculated the maximum tidal head across the barrier around both ocean high and ocean low tide (Fig. 6b). Instantaneous tidal head across the barrier results from reduction in tidal range inside the barrier, longer phase lags, and mean water level set-up as the barrier elongates to the south. The results indicate that during most of the lengthening of Nauset Beach, the head across the barrier was significantly larger near ocean low tide than near ocean high tide. The ocean low tide head also developed more rapidly during the elongation of the barrier. According to model results, low tide head reached its peak pre-breach value approximately twenty years before the breach occurred and remained relatively constant until the breach. High tide head, in contrast, continued to increase until the time of the breach.

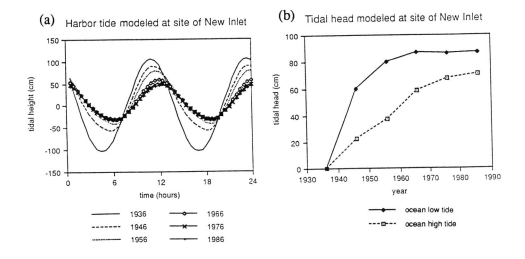

Figure 6. Model results at the future site of New Inlet for various stages during the evolution of the Chatham Harbor-Pleasant Bay Estuary: (a) tidal elevations within the estuary and (b) tidal head across the barrier. See Figure 5 for model configurations and sampling location.

These results suggest that tidal inlet formation through Nauset Beach requires — in addition to the initiating storm wave overwash — critical tidal heads across the barrier at ocean high as well as ocean low tide. Nauset Beach has been subject to numerous washovers since the turn of the century, for example during the severe winter storm of 6 February 1978 which breached Monomoy Island. Yet these earlier washovers of Nauset Beach failed to develop into permanent inlets — even though the head across the barrier at low tide already may have attained its maximum pre-breach value. Our model results suggest that the initial stage of inlet formation, when the overwash channel is deepened by strong flood currents, is crucial to permanent inlet formation. Only if and when the channel becomes deeply scoured by strong flood currents, can the even stronger ebb currents complete the formation process.

Thus it appears that the time interval between episodes of inlet formation depends ultimately on the development of a critical ocean high tide head accompanied by adequate ocean low tide gradient. This critical head results from distortion of the tidal elevation signal within the estuary at the site of the potential inlet. At Chatham, the distortion develops in response to physical changes in the form of the barrier and estuary, specifically the elongation of Nauset Beach and the formation of West Channel.

The behavior of the incipient inlet during the early days of January, 1987, supports our hypothesis of a critical high tide head. The breaching began with an overwash channel produced at perigean spring high tide during a severe easterly storm. For the first several days following that initial breaching, the new channel — while presumably deepening with each successive flood tide — was not deep enough to permit appreciable ebb flow, and across-barrier sediment transport was largely westward (see Fig. 1). It appears that only after sufficient channel deepening had been produced by flood currents, were the ebbs — driven by even greater heads — able to complete formation of New Inlet.

Of course other factors contribute to the breaching. Storm surge is required to raise mean water level in both the ocean and lagoon, so that the barrier can be overwashed and the hydraulic link can be established. This initial superelevation appears critical to the process to allow the pressure gradients to work. In addition, wave set-up on the ocean side of the barrier acts to increase the sea-to-land gradient. Wave set-up of nearly one meter may accompany five meter waves, for example. Morphologic features may also facilitate the breaching process. A weakness (blow-out) in the barrier dunes due to previous overwashes may channelize water across the barrier during storms. The likelihood of breaching may be enhanced where the barrier has been narrowed by long-term beach erosion or where the bay immediately adjacent to the barrier is unusually deep. Finally, groundwater behavior within the barrier may also play a role (e.g., Ogden, 1974; Weidman et al., in prep.).

Examination of Multiple-Inlet Stability

In order to examine the stability of the multiple-inlet system following the formation of New Inlet, we applied our model to the 1986 and 1988 configurations of the system (last two diagrams in Fig. 5). A comparison of the model results indicates that formation of New Inlet altered the fundamental tidal flow pattern through much of the southern portion of the Chatham Harbor-Pleasant Bay estuary (Fig. 7). Through West Channel, for example, modeled discharge near low water reversed direction after the development of New Inlet and discharge near high water increased markedly. Yet the most drastic changes in modeled discharge after the formation of New Inlet occurred in Middle Channel. After the breach of Nauset Beach, discharge through Middle Channel reversed direction relative to pre-breach conditions and decreased dramatically in magnitude (Fig. 8). The large decrease in modeled discharge through Middle Channel was a direct result of a drastically reduced surface gradient (Fig. 7). Before the breach, the maximum change in surface elevation over the length of Middle Channel was 20 to 30 cm. After the breach it was less than 10 cm.

Thus we conclude that formation of New Inlet produced a condition of hydrodynamical instability within the multiple-inlet system, leading to decoupling of the previously existing inlets from northern Chatham Harbor and Pleasant Bay, and the establishment of New Inlet as the primary — and only stable — channel connecting them with the open sea. The practical importance of this result for harbor management is clear: new channels that developed following the formation of New Inlet have the potential of serving

as reliable routes for navigation, but former channels that are now hydrodynamically inactive and shoaling, may not be reliable waterways and efforts to
keep their entrances open through dredging may be ineffective.

Extreme shoaling at the northern end of Middle Channel is evident from aerial
photography taken following the formation of New Inlet and is discussed
elsewhere in this volume (e.g., Liu et al.). Prior to completion of this model
study, it was thought by some that the observed shoaling might indicate that
the extremely energetic sedimentation processes associated with wave action
at New Inlet was responsible for overwhelming and altering the previously
existing hydrodynamical system at the north end of Middle Channel. However, the present results indicate that the hydrodynamical changes resulted

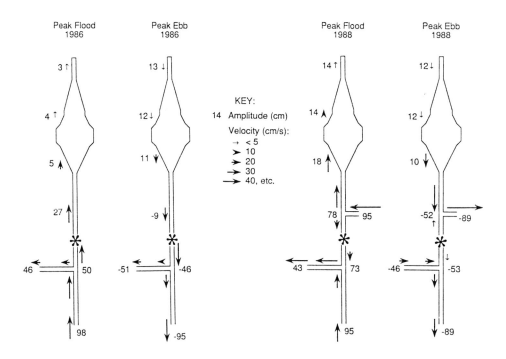

Figure 7. Model results for tidal elevation and velocity before and after the formation of New Inlet.
Values shown are concurrent with the time of peak ebb or peak flood at the site of New Inlet. Also
indicated is the site of tidal velocity measurements displayed in Figure 8.

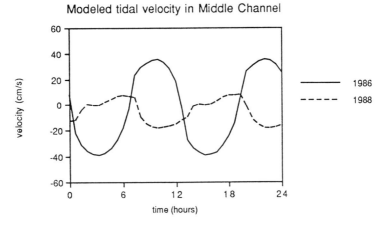

Figure 8. Model tidal velocities within Middle Channel before and after the formation of New Inlet. Positive velocities are northward. See Figure 7 for model configurations and sampling location.

primarily from the reduced Chatham Harbor-Pleasant Bay surface gradient produced by the opening of New Inlet, and that the shoaling of the northern and southern ends of Middle Channel is primarily the result, rather than the cause, of the altered hydrodynamics.

Summary

This paper addresses two specific questions concerning the role of tidal hydrodynamics in determining the long-term morphologic evolution of the Nauset Beach-Monomoy Island barrier system and the Chatham Harbor-Pleasant Bay tidal estuary: (1) why do the barrier system and tidal estuary exhibit a long-term cycle of new inlet formation, and (2) once a new inlet forms, why does the channel to the south of the inlet become unstable, resulting in the decoupling of the northern and southern sections of the estuary?

To answer these questions, we adapted an existing one-dimensional finite difference model for use in a multiple inlet system. Comparison to field observations indicates that a relatively simple model captures the fundamental linear and nonlinear hydrodynamics observed throughout the Chatham Harbor-Pleasant Bay estuary in 1988, approximately 15 months after the breach occurred.

To address question (1) specifically, the numerical model was used to recreate the basic flow patterns in the tidal estuary at ten-year intervals during the last half century. Modeling results suggest that an inlet will form through Nauset Beach when southerly elongation of the barrier has led to a critical shore-directed head across the barrier at ocean high tide. Only if a critical head exists at high tide during consecutive tidal cycles can flood currents deepen the overwash channel sufficiently to enable the stronger ebb currents to complete the formation process. Other factors, such as storm surge, wave set-up, pre-existing barrier geometry, and perhaps groundwater behavior, can contribute to the breaching.

To address question (2), the model was used to compare the flow conditions in 1986 and 1988, shortly before and shortly after the formation of the new inlet. Model results indicate that formation of the inlet drastically reduces the surface gradient along the pre-existing channel to the south of the inlet. The reduced surface gradient in turn reduces the discharge and velocity through the channel, eliminating the tidal scouring action needed to keep the channel entrances open. The observed accumulation of sand marks the final phase of tidal decoupling.

Model trends suggest the inlet formation potential at other sites may be described by similar studies. Other areas of Cape Cod, such as Nauset Inlet to the north and Popponesset Inlet at Mashpee, experience barrier spit elongation and hence changes in tidal properties within embayments served by inlets.

Acknowledgments

This work is the result of research sponsored by NOAA National Sea Grant College Program Office, Department of Commerce, under Grant No. NA88-AA-D-SG090, Woods Hole Oceanographic Institution Sea Grant Project No. R/O-6. The U.S. Government is authorized to produce and distribute reprints for governmental purposes notwithstanding any copyright notation that may appear hereon. Support for this work also was provided by the U.S. Army Corps of Engineers (New England Division), the Town of Chatham, and the W.H.O.I. Coastal Research Center. Woods Hole Oceanographic Institution Contribution No. 8317.

References

Aubrey, D. G., 1986. A study of bluff erosion at Morris Island, Chatham, MA. Aubrey Consulting, Inc., A report submitted to local residents, Falmouth, MA, 55 p. + appendices.

Aubrey, D. G., and P. E. Speer, 1984. Updrift migration of tidal inlets. *Journal of Geology*, v. 92, p. 531-545.

FitzGerald, D. M., and S. A. FitzGerald, 1977. Factors influencing tidal inlet throat geometry. In: Coastal Sediments 1977. *American Society of Civil Engineers,* New York, p. 563-581.

Friedrichs, C. T., and D. G. Aubrey, 1988. Non-linear tidal distortion in shallow well-mixed estuaries: a synthesis. *Estuarine, Coastal and Shelf Science,* v. 27:, p. 521-545.

Friedrichs, C. T., and D. G. Aubrey, 1989. Numerical modeling of Nauset Inlet/Marsh. In: C. T. Roman and K. W. Able (eds.), *An ecological analysis of Nauset Marsh,* Center for Coastal and Environmental Studies, Rutgers - The State University of New Jersey, New Brunswick, NJ, Appendix D, p. A179-A222.

Friedrichs, C. T., and O. S. Madsen, 1992. Nonlinear diffusion of the tidal signal in frictionally dominated embayments. *Journal of Geophysical Research,* v. 97, p. 5637-5650.

Giese, G. S., 1978. The barrier beaches of Chatham, Massachusetts. Provincetown Center for Coastal Studies, April 1978 Report, Provincetown, MA, 7 pp.

Giese, G. S., 1988. Cyclical behavior of the tidal inlet at Nauset Beach, Chatham, Massachusetts. In: D. G. Aubrey and L. Weishar (eds.), *Hydrodynamics and Sediment Dynamics of Tidal Inlets, Lecture Notes on Coastal and Estuarine Studies,* v. 29, Springer-Verlag, New York, p. 269-281.

Giese, G. S., D. G. Aubrey and J. T. Liu, 1989. Development, characteristics, and effects of the new Chatham Harbor inlet. Woods Hole Oceanographic Institution, WHOI-89-19, Woods Hole, MA, 32 pp.

Giese, G. S., C. T. Friedrichs, D. G. Aubrey and R. G. Lewis II, 1990. Application and assessment of a shallow-water tide model to Pamet River, Truro, Massachusetts. A report submitted to the Truro Conservation Trust, Truro, MA, 26 pp.

Goldsmith, V., 1972. Coastal processes of a barrier island complex and adjacent ocean floor: Monomoy Island-Nauset Spit, Cape Cod, Massachusetts. Ph.D. thesis, University of Massachusetts, Amherst, MA, 469 pp.

Leatherman, S. P., and R. E. Zaremba, 1986. Dynamics of a northern barrier beach: Nauset Spit, Cape Cod, Massachusetts. *Geological Society of America Bulletin,* v. 97, p. 116-124.

Liu, J. T., D. K. Stauble, G. S. Giese and D. G. Aubrey, this volume. Morphodynamic evolution of a newly formed tidal inlet. In: Aubrey, D.G. and G. S. Giese (eds.), *Formation and Evolution of Multiple Inlet Systems. Coastal and Estuarine Studies Series,* AGU.

McClennen, C. E., 1979. Nauset Spit: model of cyclical breaching and spit regeneration during coastal retreat. In: S. P. Leatherman (ed.), *Environmental Geological Guide to Cape Cod National Seashore,* S.E.P.M., Eastern Section Field Trip Guide Book. Boulder, CO., p. 109-118.

Mitchell, H., 1874. Report to Prof. Benjamin Pierce, Superintendent United States Coast Survey, concerning Nauset Beach and the peninsula of Monomoy. In: *Report of the Superintendent of the United States Coast Survey for 1871*, Appendix No. 9, p. 134-143.

Ogden, J. G., 1974. Shoreline changes along the Southeastern coast of Martha's Vineyard, Massachusetts for the past 200 years. *Quaternary Research,* v. 4, p. 496-508.

Oldale, R. N., J. D. Friedman and R. S. Williams Jr., 1971. Changes in coastal morphology of Monomoy Island, Cape Cod, Massachusetts. U.S. Geological Survey Professional Paper, v. 750-B, p. B101-B107.

Speer, P. E., and D. G. Aubrey, 1985. A study of non-linear tidal propagation in shallow inlet/ estuarine systems, part II: theory. *Estuarine, Coastal and Shelf Science,* v. 21, p. 207-224.

U.S. Army Corps of Engineers, 1968. Survey report: Pleasant Bay, Chatham, Orleans, Harwich, Massachusetts. Department of the Army, New England Division, Corps of Engineers, Waltham, MA, 61 p. + appendices.

Weidman, C. R., D. G. Aubrey and C. T. Friedrichs, in prep. Tidal dynamics of the water table in a barrier beach.

4

Tidal Residual Currents and Sediment Transport Through Multiple Tidal Inlets

James T. Liu and David G. Aubrey

Abstract

Tidal residual currents in a tidal channel connecting two water bodies having contrasting tides are most sensitive to the mean sea-level differences, less sensitive to the tidal amplitude differences, and least sensitive to the tidal phase differences between the two ends of the channel. On the other hand, tidal phase difference is the most important factor in generating M_4 overtide in the channel, the tidal amplitude difference has intermediate impact on M_4 generation, and the mean sea-level difference has little effect on M_4 generation. Residual currents and M_4 overtide are generated by different mechanisms. The former is related to the friction in the system, and the latter is related to the kinematic non-linearity in the system. The combined influence of the tidal phase, amplitude, and mean sea-level differences on the generation of tidal residual currents in a channel is complex and non-linear, and can be predicted properly only by non-linear numerical models. The sediment transport patterns in a tidal channel that connects two bodies of water having different tidal characteristics can be attributed to residual currents that is primarily caused by the mean sea-level difference, and to a minor degree, by the tidal amplitude and phase differences between the two ends of the channel. Multiple inlets at Chatham, Massachusetts, are used as a case study.

Formation and Evolution of Multiple Tidal Inlets
Coastal and Estuarine Studies, Volume 44, Pages 113-157

Introduction

The tidal-mean exchange of salt, pollutants, suspended material, and sediments through a coastal channel respond to residual currents. In weakly-to-strongly non-linear estuaries, bays, and channels, the dominant tide is the primary generator of residual currents (Cotter, 1974; Tee, 1976, 1977; van de Kreeke, 1978, 1980; Ianniello, 1977, 1979; Huang et al., 1986; Wong, 1989).

Tidally generated residual currents in a channel have been shown to be influenced by the tidal amplitude, phase, and mean sea-level differences between the two open boundaries (van de Kreeke, 1980; Huang et al., 1986; Wong, 1989), and by the geometric asymmetry of the channel between the two open ends (Cotter, 1974). The present study further investigates the individual importance of the dominant tidal amplitude, phase, and mean sea-level difference on the tidal characteristics and the generation of residual currents in the interior of an open channel through diagnostic numerical modeling exercises. The model findings are then used to interpret field observations from a tidal channel that connects two bodies of water having different tidal characteristics. The effect of the residual currents on the long-term sediment transport in that channel is deduced from historical evidence.

Study Site

Chatham Harbor is a bar-built, multiple-inlet system located at the southeastern corner of Cape Cod, Massachusetts (Fig. 1). The barrier spit (Nauset Beach) that separates the estuary from the Atlantic Ocean had been accreting southward since the late 1800's until January, 1987, when it was breached during a severe storm to form a new tidal inlet (Giese, 1988). West Channel (a tidal inlet), the focus of this study, separates Monomoy Island from Morris Island, linking Nantucket Sound indirectly with the Atlantic Ocean through the southern part of the estuary (Fig. 1).

West Channel was formed in 1957 when Monomoy Island detached from Morris Island. From the early 1960's to the early 1970's before the southern tip of Nauset Beach overlapped with Monomoy Island, West Channel was directly under the influence of the coastal waves and longshore currents of the

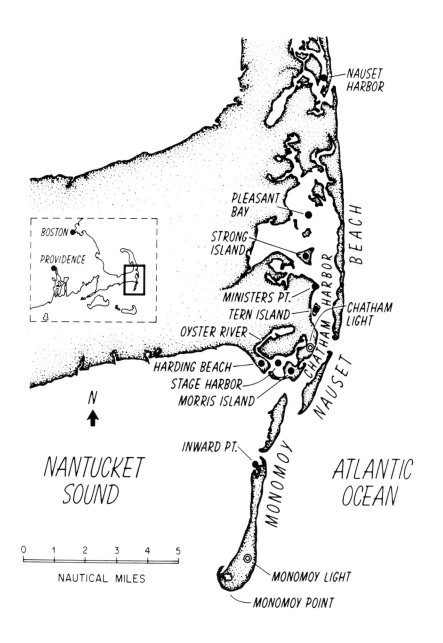

Figure 1. Index map for Chatham Harbor, Massachusetts.

Atlantic Ocean. Documentation of sand bodies and associated bedforms on either side of West Channel and measurements of tidal currents on the Nantucket side of the inlet in the early 1970's (Hine, 1975) suggested a dominant sediment transport from the Atlantic Ocean into Nantucket Sound. About 1975, Nauset Beach extended south of West Channel, gradually sheltering this inlet from the direct impact of Atlantic Ocean waves.

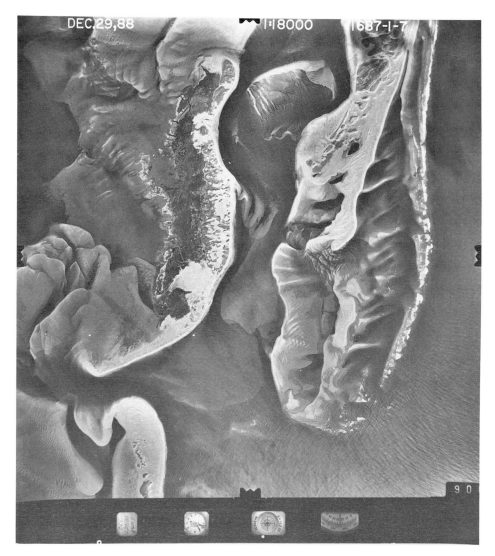

Figure 2. Vertical aerial photograph of the breach on Monomoy Island (lower left corner) taken in December, 1988. The Atlantic Ocean is on the right hand side of the breach, and Nantucket Sound is on the left.

In 1978, Monomoy Island was breached approximately 3.6 km south of West Channel (Fig. 2). Since then, extensive sand bodies have enlarged on the Nantucket side of the breach, resembling a flood tidal delta-complex consisting of superimposing spillover lobes. Little sand has accumulated on the Atlantic side of the breach (Fig. 2). This imbalance of sediment deposition between the two sides of the breach suggests a net sediment transport from the Atlantic Ocean to Nantucket Sound (Liu et al., 1989), though greater longshore sand transport on the Atlantic side undoubtedly plays some role.

Recent aerial photographs taken between May and December, 1988, document shoal development and sand flat growth on the Nantucket side of West Channel, indicating a dominant westward sediment transport following the recent breach of Nauset Beach (Liu et al., 1989). Because of the consistency in sediment transport dominance observed in West Channel and the breach on Monomoy Island, we hypothesize that tidal elevation differences between Nantucket Sound and the Atlantic Ocean generate residual flows controlling sediment transport.

In April, 1988, two digitally-recording tide gauges were deployed for 29 days in the southern part of Chatham Harbor: one due east of West Channel, and one in Nantucket Sound near the western end of the inlet (Fig. 3). During the same period, a shorter (7-day) record of near-bed current speed and sea-surface height was obtained from the middle of West Channel approximately 750 m from the tide gauge in south Chatham Harbor, and 1250 m from the tide gauge in Nantucket Sound (Fig. 3). Prior to deployment, the electromagnetic current meter was calibrated in the laboratory to yield expected mean errors of 1 to 2 cm/sec. The bathymetry in south Chatham Harbor and West Channel was surveyed using an integrated navigation system.

Thirty-five tidal constituents were resolved for each of the two tidal records (Fig. 4; the major constituents are listed in Tables 1 and 2). The M_2 tide is the largest tide in both south Chatham Harbor and Nantucket Sound. The tidal amplitude in south Chatham Harbor is 40 cm greater than that in Nantucket Sound. The M_2 phase in Nantucket Sound leads that in south Chatham Harbor by 5.5 degrees (Tables 1 and 2).

Tidal currents in the middle of West Channel show strong westward (positive) flows having a maximum speed of approximately 120 cm/sec, which is about

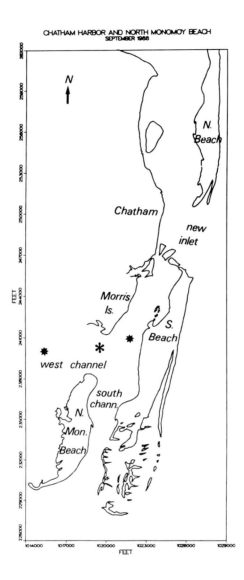

Figure 3. Detailed map showing the study area. The two asterisks represent locations of the two tide gauges deployed in south Chatham Harbor and Nantucket Sound. The middle symbol marks the location of the current meter deployment.

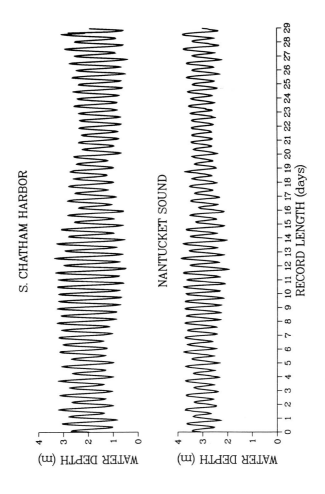

Figure 4. Tidal height records from south Chatham Harbor and Nantucket Sound.

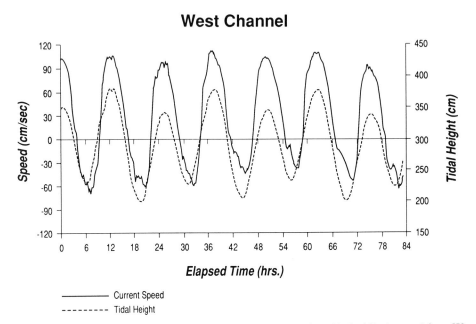

Figure 5. Instantaneous current speed (solid line) and sea-surface (dashed line) record from West

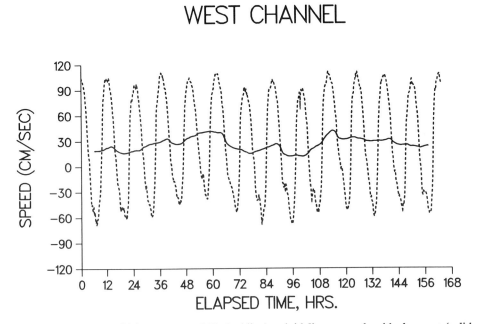

Figure 6. Instantaneous tidal current speed (dashed line) and tidally averaged residual current (solid line) in West Channel.

TABLE 1. Major Tidal Constituents in south Chatham Harbor.

Constituent	Period (hr.)	Amplitude (m)	Greenwich Phase (degrees)	% Total Energy
M_2	12.42	0.97	99.5	87.53
S_2	12.0	0.17	123.4	2.91
N_2	12.66	0.187	50.5	3.27
K_1	23.93	0.133	191.9	2.20
M_4	6.21	0.03	104.8	0.08
RMS Tidal Height = 1.98 m				

TABLE 2. Major Tidal Constituents in Nantucket Sound.

Constituent	Period (hr.)	Amplitude (m)	Greenwich Phase (degrees)	% Total Energy
M_2	12.42	0.561	94.0	82.54
S_2	12.0	0.091	119.0	2.33
N_2	12.66	0.136	43.8	4.89
K_1	23.93	0.099	175.5	3.43
M_4	6.21	0.048	296.1	0.57
RMS Tidal Height = 1.18 m				

twice the maximum of the eastward flows (Fig. 5). The phase equality between tidal current speed and sea-surface elevation suggests a progressive tide propagating through the inlet. To examine the residual velocity, the current speed record was averaged over the M_2 period (12.42 hr.) yielding a flow from south Chatham Harbor into Nantucket Sound (Fig. 6). This residual current displays some subtidal variation, about a mean speed of 26 cm/sec to the west.

Field data document flow in West Channel from greater tidal amplitude (south Chatham Harbor) to smaller tidal amplitude (Nantucket Sound). However, the contributions to residual flow from the tidal phase difference and mean sea-level difference (of which we have no direct observations) are unclear. Therefore, modeling was used to investigate the relative influences of tidal amplitude, phase, and mean sea-level difference on the generation of residual currents in a hypothetical channel to determine the relative importance of the three factors. The findings are then related to the field observations from West Channel.

Definitions and Model Formulation

The vertically averaged transport ($\bar{q}$) in a tidal channel is defined as:

$$\bar{q} = \bar{u} \, (h+\eta) \tag{1}$$

in which $\bar{u}$ (t) is the vertically averaged instantaneous Eulerian velocity, h is the still water depth, and η (t) is the sea-surface elevation. The time average over the M_2 period yields the residual flow:

$$\langle \bar{q} \rangle = \langle \bar{u}h \rangle + \langle \bar{u}\eta \rangle = \langle \bar{u} \rangle h + \langle \bar{u}\eta \rangle \tag{2}$$

The first term on the right hand side of equation 2 is directly proportional to the Eulerian mean flow:

$$U_E = \langle \bar{u} \rangle \tag{3}$$

The second term, the Stokes transport (assuming small amplitude long waves), results from the co-oscillation of the sea-surface elevation and the instantaneous velocity, $\bar{u}$. The Stokes drift is obtained by depth-averaging the second term on the right hand side of eq. 2:

$$U_S = \frac{\langle \bar{u}\eta \rangle}{h} \tag{4}$$

The depth-averaged Lagrangian velocity U_L is:

$$U_L = U_E + U_S \tag{5}$$

A straight east-west oriented (x-axis positive towards the west) channel, 2 km in length (Fig. 7), having a constant trapezoidal cross-section with tidal flats on both sides, is used to represent West Channel (Fig. 8).

The governing equations for the one-dimensional, cross-sectionally averaged flow, $Q(x,t)$, and sea-surface elevation, $H(x,t)$, in the model channel are:

Continuity: $$\frac{\partial H}{\partial t} - \frac{1}{b}\frac{\partial Q}{\partial x} = 0 \tag{6}$$

Momentum: $$\frac{\partial Q}{\partial t} + \frac{\partial}{\partial x}\left(\frac{Q^2}{A}\right) = -gA\frac{\partial H}{\partial x} - \frac{F}{A \cdot R}|Q|Q \tag{7}$$

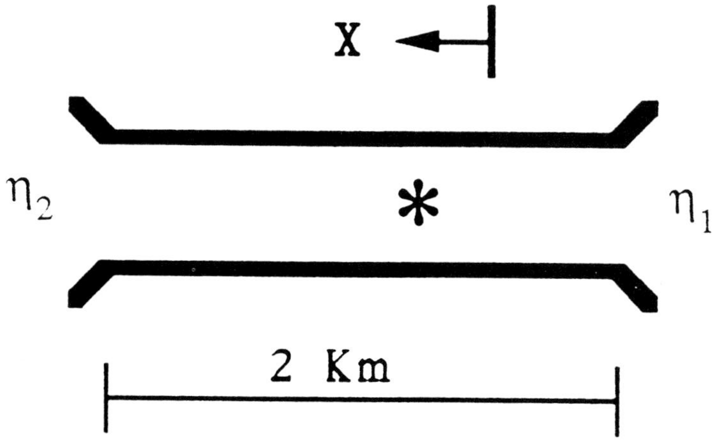

Figure 7. The hypothetical tidal channel for the model. The pond symbol marks the grid point at which tidal current and sea-surface are examined.

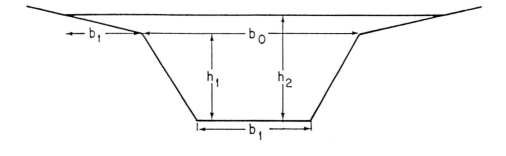

Figure 8. The geometry of the trapezoidal channel cross-section:
$\qquad$ b_1 = 200 m, width of channel at its base
$\qquad$ b_o = 350 m, width of channel at its tip
$\qquad$ b_t = 400 m, width of tidal flat
$\qquad$ h_1 = 1.85 m, channel depth
$\qquad$ h_2 = undisturbed water depth

where b is the width, A is the cross-section, R is the hydraulic radius of the channel, and F=0.02, is a friction parameter. This formulation has been shown to be suitable for strongly non-linear systems (Speer and Aubrey, 1985; Aubrey and Friedrichs, 1988). The governing equations were solved numerically, using an explicit, leap-frog finite difference scheme. The model consists of 9 grid points with spacings of 250 m, and time-steps of 15 sec.

The boundary conditions are assumed as follows:

$$\text{at } x = 0, \quad \eta_1 = a_1 \cos(\omega t)$$

$$\text{at } x = L, \quad \eta_2 = a_2 \cos(\omega t - \phi)$$

where L is the length of the channel, a_1 and a_2 are tidal amplitudes at the two open ends of the channel, ϕ is the tidal phase difference, h_o is the mean sea-level difference, and ω is the M_2 tidal frequency. a_1 is fixed to be 1 m, and the still water depth at x=0 is set to 2.5 m. During the model exercises, the values of a_1/a_2, f, and h_o are varied one at a time. The tidal characteristics produced by each model run at the grid point corresponding to the current meter deployment in West Channel are examined, and their Lagrangian, Eulerian, and Stokes velocities and transports are calculated.

Model Results

Influence of the Tidal Amplitude Difference

The model was run for 7 cases in which a_1/a_2 was set to be 0.5, 0.75, 1.0, 1.25, 1.5, 2.0, and 2.5, respectively, while ϕ and h_o were set to zero. When the tidal amplitude at x=L is greater ($a_1/a_2 < 0$), the instantaneous tidal current speed and sea-surface elevation in the interior of the channel are in quadrature (Fig. 9a). The tidal current shows easterly dominance as indicated by the greater negative speed and sharper gradient around the time of maximum water elevation (Fig. 9a). As a result of this flow dominance, the Eulerian, Stokes, and Lagrangian residual currents and transport are directed towards the east (Figs. 9b,c).

When $a_1/a_2 = 1$, the sea-surface in the interior of the channel simply co-oscillates with the tides at the two ends, and there is little current in the channel (Fig. 9a). The result is no residual current or transport through the channel (Figs. 9b,c).

When a_1/a_2 becomes greater than unity, the instantaneous tidal current and sea-surface elevation in the interior of the channel are in phase. The flow shows a westward dominance. As a_1/a_2 increases, so does the amplitude of the instantaneous tidal flow and the asymmetry between the flood and ebb currents (Figs. 9a,c), a condition that results in the increase of westward residual velocities and transports (Figs. 9b, c).

The sensitivity analysis of residual current-generation to the tidal amplitude differences shows that the residual current curves become asymmetrical about unity (Fig. 9d). Since the value of a_1 is fixed at 1 m, as a_1/a_2 increases beyond unity, a_2 becomes smaller, which results in the asymptotically slow increase of residual velocities. On the other hand, when a_1/a_2 decreases below unity (a_2 greater than 1 m), the magnitude of residual currents increases quickly due to the approach of a_2 to the mean water depth (h=2.5 m), causing stronger non-linearity in the system. The boundary tidal amplitude differences do not affect the characteristics of the tide in the interior of the channel, except that the phasing between the instantaneous tidal current and sea-surface elevation varies between zero and 180 degrees depending on which end of the channel has greater tidal amplitude.

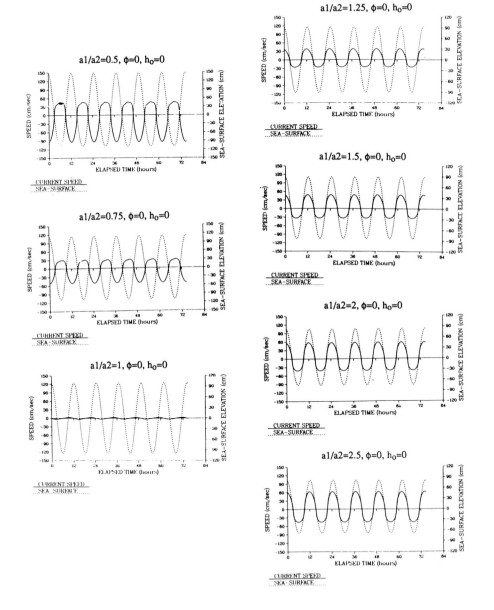

Figure 9. Influence of tidal amplitude differences:
(a) Instantaneous tidal current speed and sea-surface

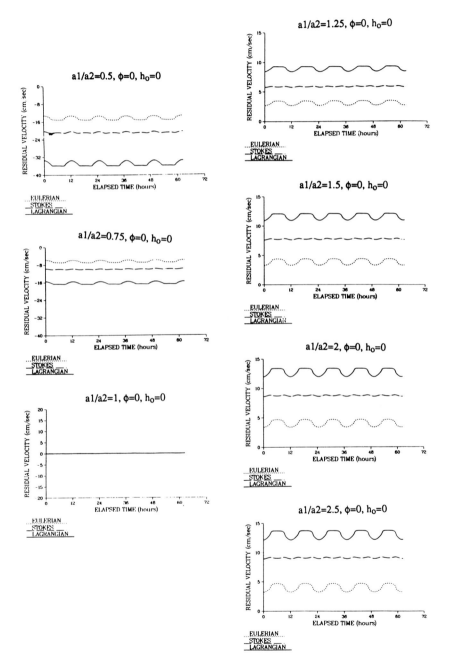

(b) Residual currents

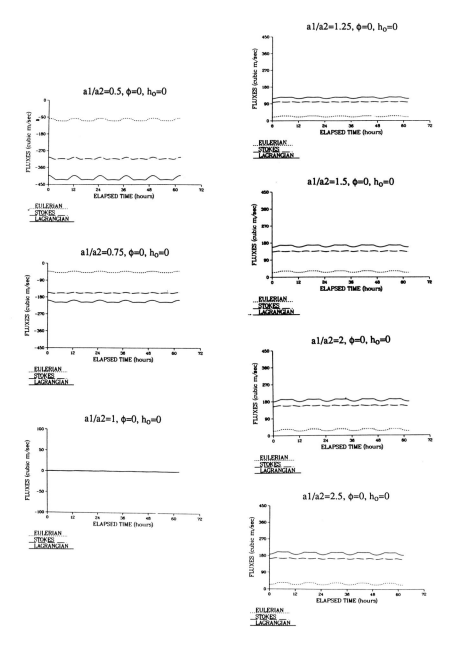

(c) Residual transports

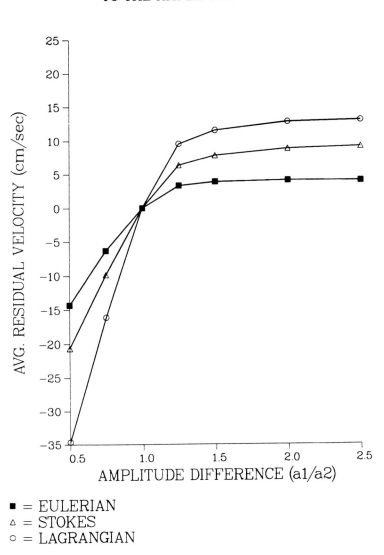

(d) Residual currents versus tidal amplitude differences

Influence of the Tidal Phase Difference

The influence of tidal phase difference was tested for $\phi = -90, -60, -30, 30, 60,$ and 90 degrees. In cases of negative phase differences when the phase at x=L leads that at x=0, the sea-surface elevation leads the instantaneous tidal flow in the interior of the channel (Fig. 10a). The Stokes drift is eastward, whereas the Eulerian and Lagrangian velocities are in the opposite direction (Fig. 10b). When ϕ is near -90 degrees, the Stokes transport dominates the opposing Eulerian transport, so that the resultant Lagrangian transport is in the direction of the Stokes transport (Fig. 10c). As ϕ tends towards zero, Eulerian transport becomes dominant, and the Lagrangian transport is in the direction of the Eulerian transport (Fig. 10c).

In cases of positive phase differences, the sea-surface lags behind the instantaneous tidal flow in the interior of the channel (Fig. 10a). The Stokes velocity remains positive (westwards) in all cases, but the Eulerian velocity changes direction from westward to eastward as ϕ increases towards 90 degrees (Figs. 10b,c). The Stokes transport dominates when ϕ is positive, so that the resultant Lagrangian transport is in the same direction as the Stokes transport (Fig. 10c).

The tidal phase difference appears not to change the direction of the Stokes velocity, which remains westwards in all cases. For negative ϕ's, the Eulerian flow is constantly easterly (negative). But for positive ϕ's, the direction of the Eulerian flow is variable (Fig. 10d).

Influence of the Mean Sea-level Difference

The influence of the mean sea-level difference was tested for $h_o = -0.1, -0.05, -0.01, 0.01, 0.05,$ and 0.1 m. A mean sea-level difference creates a noticeable mean flow (Fig. 11a). In addition, the amplitude of the tidal flow is suppressed.

Physically the change of h_o is analogous to tilting the sea-level along the channel, having the pivotal point at x=0. Negative h_o creates a westward sloping sea-surface, thereby resulting in westerly residual currents (Figs. 11a,

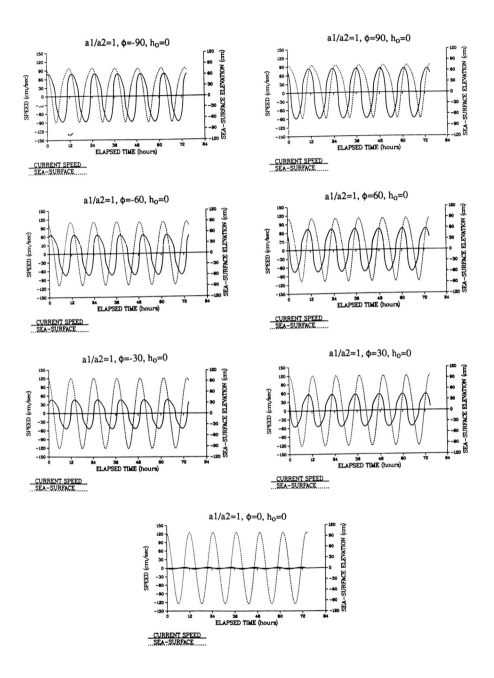

Figure 10. Influence of tidal phase differences:
(a) Instantaneous tidal current speed and sea-surface

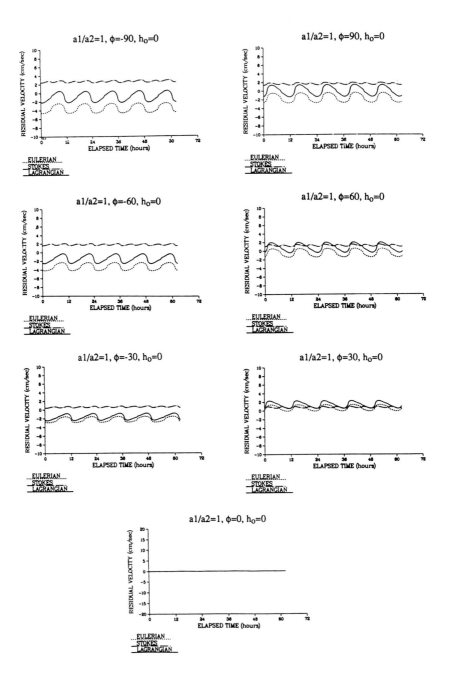

(b) Residual currents

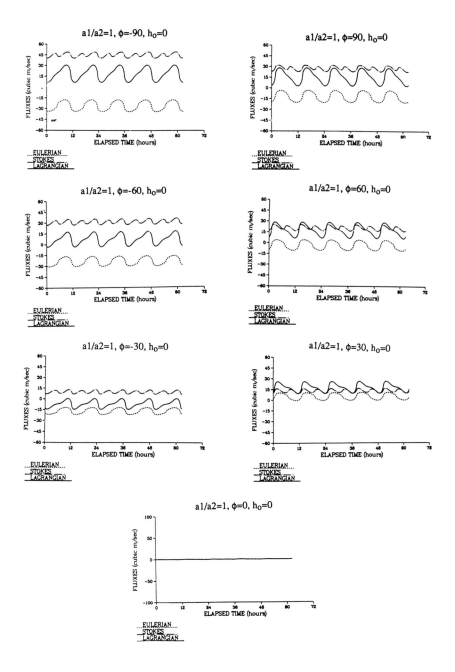

(c) Residual transports

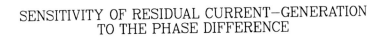

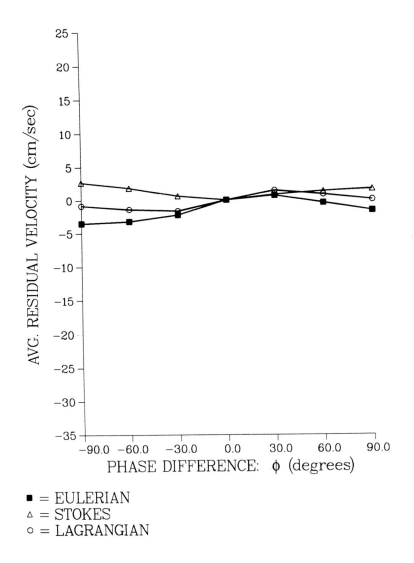

(d) Residual currents versus tidal phase differences

b). On the other hand, a positive h_o results in easterly flows. Within the suppressed instantaneous tidal flow in the interior of the channel, there is a tidal asymmetry in the presence of a residual flow (Fig. 11a). However, as the absolute value of h_o increases, this tidal asymmetry in the instantaneous flow gradually disappears. In general, for corresponding h_o of opposite signs, the residual velocities and their transports are identical in magnitude, but are rotated 180 degrees and symmetrical around the 0-axis (Figs. 11a, b, c).

In the case of mean sea-level difference, the Stokes velocity and transport are fairly insignificant, compared to the Eulerian velocity and transport (Figs. 11c, d). Due to the shortness of the channel (2 km long), the system is sensitive to the difference in the mean sea-level. Numerical model results indicate that even a slight h_o of 1 cm in either direction can produce a pressure gradient on the order of 10^{-5}, which can result in residual Eulerian flows of 5 cm/sec.

Discussion

The Generation of M_4 Overtide

The generation of residual currents in a tidal channel is most sensitive to the mean sea-level difference (Fig. 11d), less sensitive to the tidal amplitude difference (Fig. 9d), and least sensitive to the tidal phase difference (Fig. 10d) between the two ends. One the other hand, the Eulerian residual currents produced by the tidal phase differences show distinctive modulations at semi-diurnal frequencies (Fig. 10b). Eulerian residual currents produced by tidal amplitude differences show slight semi-diurnal modulations (Fig. 9b), and those produced by the mean sea-level differences have little modulation (Fig. 11b). These modulations might have been introduced by the time-averaging process to obtain residuals, and therefore, probably do not reflect the physics. In order to understand the discrepancies among the characteristics of the instantaneous and residual currents generated by different boundary conditions, it is worthwhile to investigate the generation of the M_4 overtide (the first harmonic of M_2) in the interior of the channel associated with the changing boundary conditions, since both the M_4 overtide and residual currents are related to the non-linearity in the system (Pingree and Maddock, 1977; Prandle, 1978).

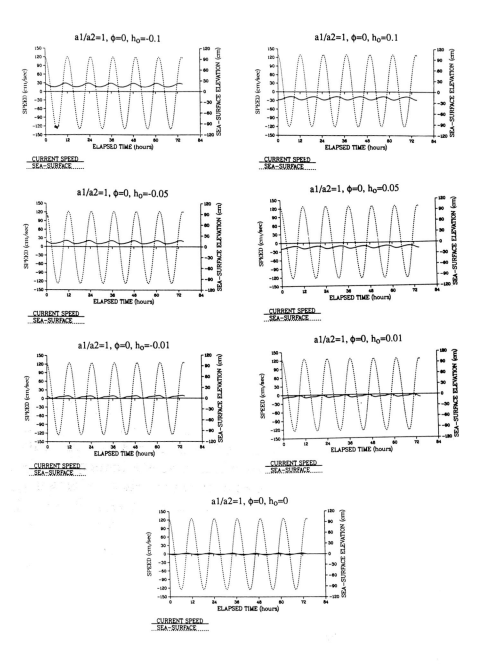

Figure 11. Influence of mean sea-level differences:
(a) Instantaneous tidal current speed and sea-surface

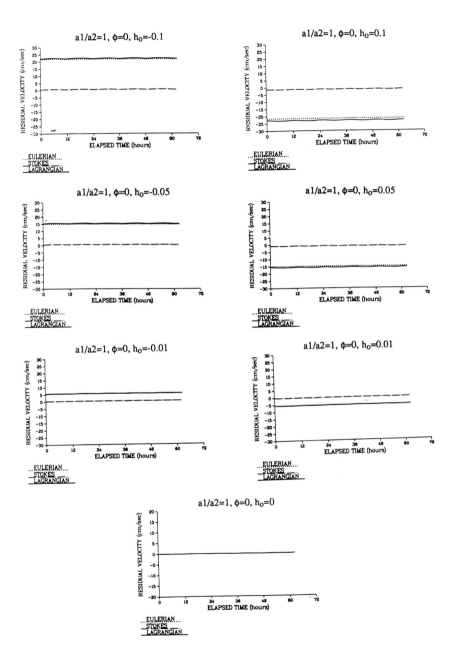

(b) Residual currents

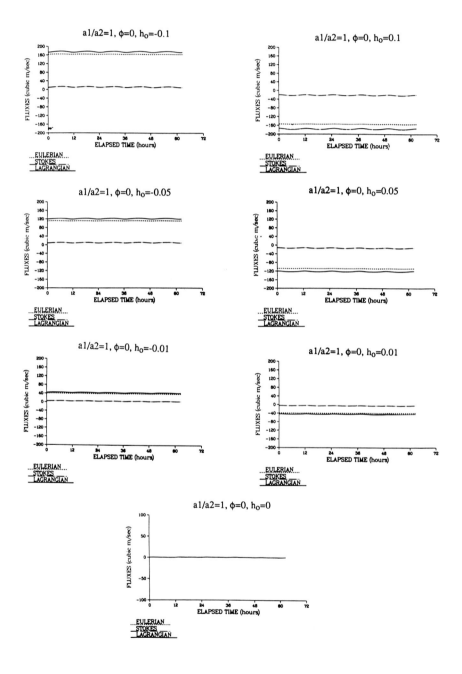

(c) Residual transports

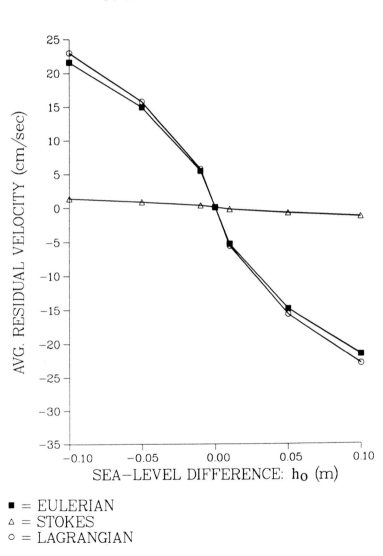

(d) Residual currents versus mean sea-level differences

The M_4/M_2 ratio, which is an indication of non-linearity, and the M_4 relative phase are plotted against the model variables (Figs. 12a, b, and c). The M_4 relative phase is defined as $\theta_{M4} = 2\,\phi_{M2} - \phi_{M4}$, where ϕ_{M2} and ϕ_{M4} are M_2 and M_4 phases respectively. M_4/M_2 is most sensitive to the tidal phase differences, less sensitive to the tidal amplitude differences, and least sensitive to the mean sea-level differences. Sea-level differences tend to suppress the generation of overtides, and the tidal amplitude differences have intermediate influence on the generation of overtides. The tidal amplitude differences have moderate effects on the non-linearity of the instantaneous and residual currents.

Friction Effects on Model Results

The distinctions between residual current and M_4 generation suggest different mechanisms to generate residual currents and overtides. Since in this particular system, the quadratic friction is the dominant non-linear term, it is helpful to examine it closely. Heath (1980) expressed the quadratic friction in his depth-integrated equation as:

$$\frac{r}{h + \zeta}\,\left| u_1 + u_2 + u_0 \right|\,(u_1 + u_2 + u_0) \tag{8}$$

where r is the quadratic friction coefficient, h is the undisturbed water depth, z is the sea-surface elevation, u_1 and u_2 are M_2 and M_4 speeds respectively, and u_0 is the residual current. Through expansion of eq. 8, Heath identified four major terms in the friction as the production of M_2 tidal energy, the input to M_4 tide, the frictional dissipation of M_4, and the input to the mean flow. Although Heath (1980) did not address the issue of the generation of M_4 and residual currents, his analysis does give a clear picture of different frictional contributions to the M_4 overtide and the residual flow. When mean sea-level differences exist, the residual flow generation overwhelms the harmonic generation, and *vice versa* in the case of tidal phase differences. At this point, the physics of the M_4 generation, its relative phase, and residual currents are not clear.

To examine further the friction effects on model results, two additional values of 0.1 and 0.005 were used for the friction parameter in the model. Curves of M_4/M_2 and M_4 relative phase versus model variables having different friction parameters show that they are identical, except for some variation in the M_4

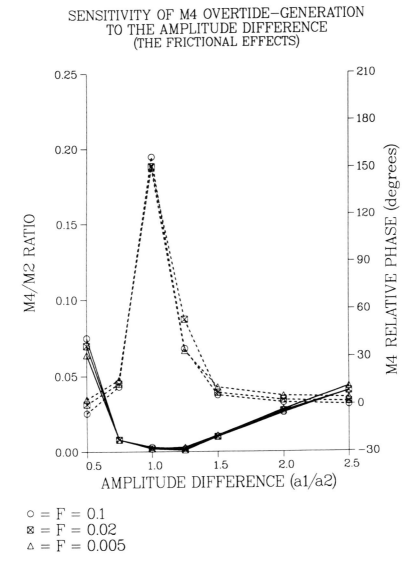

Figure 12. M_4/M_2 (solid lines) and M_4 relative phase (dashed lines) versus:
(a) Tidal amplitude differences

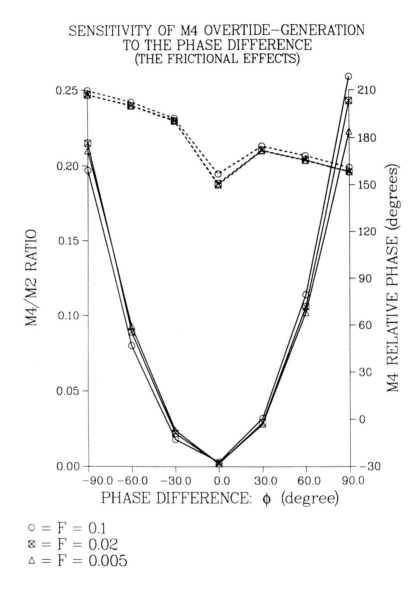

(b) Tidal phase differences

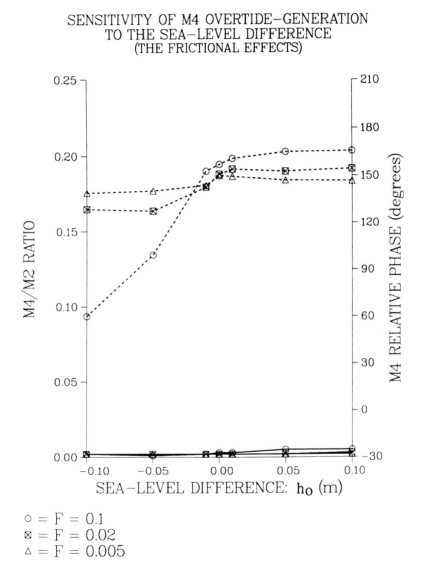

(c) Mean sea-level differences with 3 different friction parameters

relative phase in the case of mean sea-level differences (Figs. 12a, b and c). The resemblance among these curves indicate that the generation of M_4 overtides is little affected by the friction. In other words, overtides are most likely to be related to the kinematic non-linearity that entered through continuity, which means the geometry of the channel plays an important role in introducing overtides to the system.

On the other hand, higher friction in the system does reduce the magnitude of Lagrangian residual currents and *vice versa* (Figs. 13a, b and c). The friction effect is greatest in the case of mean sea-level differences (Fig. 13c), intermediate in the case of amplitude differences (Fig. 13a), and the smallest in the case of phase differences (Fig. 13b). The different sensitivity of the generation of residual currents and overtides to the friction further indicates that different mechanisms are responsible for the generation of the two in a non-linear system.

Comparisons of Model Results with Field Data

Harmonic analysis of the tidal records from south Chatham Harbor and Nantucket Sound indicates that $a_1/a_2 = 1.7$ and the phase difference is $\phi=5.5$ degrees for the M_2 tide between the two ends of West Channel (Tables 1 and 2). The diagnostic model indicates a tidal amplitude difference of $a_1/a_2 = 1.7$ only can produce a westerly Eulerian residual current of approximately 5 cm/ sec. The contribution from the tidal phase difference of $\phi = 5.5$ degrees is almost undetectable. In order to produce a model residual current having a magnitude comparable to that is observed in the field, a mean sea-level difference between south Chatham Harbor and Nantucket Sound is needed; this pressure gradient has to be from south Chatham Harbor down towards Nantucket Sound. Subsequently, trial model runs were conducted, using $a_1/a_2 = 1.7$, $\phi = 5.5°$, and varying h_o ($h_o < 0$). The value $h_o = -15$ cm generates an instantaneous tidal current that approximates the observed flow, where the maximum westward current magnitude is approximately twice that of the eastward flow (Figs. 14, 5). This finding suggests that the mean sea-level in south Chatham Harbor may be roughly 15 cm higher than that of Nantucket Sound. However, our field observations were not accurate enough to verify this difference in mean water levels.

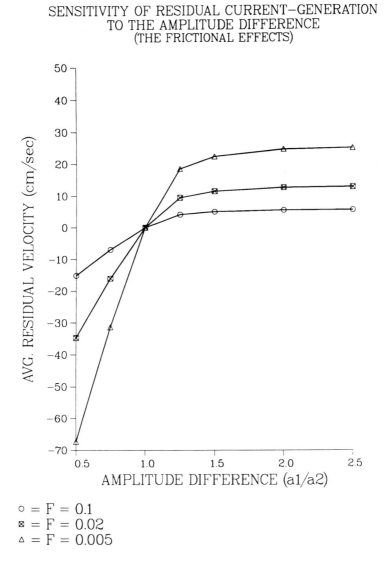

Figure 13. Friction effects on the generation of residual currents in cases of
(a) Tidal amplitude differences

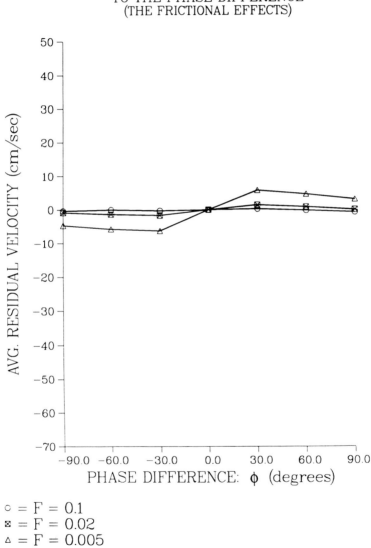

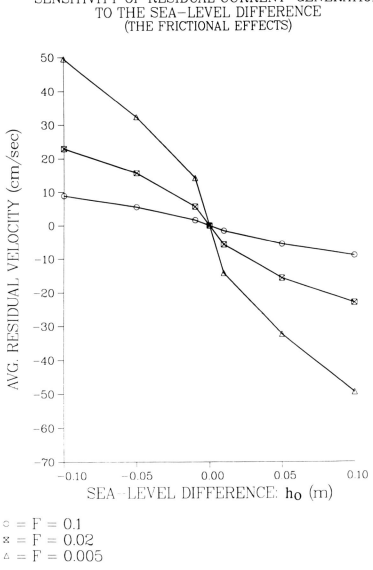

(c) Mean sea-level differences

Since the tidal currents were measured at a point, and the model simulation produces sectionally averaged flows, no direct comparison can be made between the two without invoking some additional constraints. In order to compare the field observation and the model simulations, sectionally averaged 'field data' were generated by applying the measured sea-surface in south Chatham Harbor and Nantucket Sound and a mean sea-level difference of - 15 cm. The spectral results more closely approximate the field data (Fig. 15). The residual currents from the 'real-time' model show fluctuations at tidal and subtidal frequencies (Fig. 15). The tidal frequencies indicate incomplete filtering in our analysis, whereas the subtidal frequencies may represent non-linear physics. The mean value for the Lagrangian transport of the 'real-time' model (Fig. 15) is approximately the same as that of the line spectral (single frequency) model (Fig. 14). Therefore, that the mean sea-level in south Chatham Harbor is approximately 15 cm higher than that in Nantucket Sound appears consistent.

Model Comparison

To understand further the combined effects of tidal amplitude, phase, and mean sea-level differences on the generation of residual currents, the numerical model was modified to have a rectangular channel; these results were compared with an analytical model developed by van de Kreeke (1980), having linearized friction and rectangular channel cross-section. The solution of van de Kreeke's model for tidally-averaged flow, $\bar{q}$, in the middle of the channel is expressed as:

$$\frac{\bar{q}T}{hL} = P \frac{a_1 a_2}{h^2} \sin(\phi) + Q \frac{a_1^2 - a_2^2}{h^2} + \frac{C_o h}{F_1 L} \frac{\lambda}{L} \frac{h_o}{h} \qquad (9)$$

in which h is the undisturbed water depth, T is the M_2 period, L is half the channel length, λ is the tidal wave length for zero friction, $C_o = \lambda/T$, F_1 is the linearized friction, and P and Q are functions of $(C_o h) / (F_1 L)$ and λ/T. The values for P and Q were extrapolated from tables in van de Kreeke and Dean (1975) to be 75 and 12, respectively. The three terms on the right hand side of eq. 9 are the contributions from the tidal phase, amplitude, and mean sea-

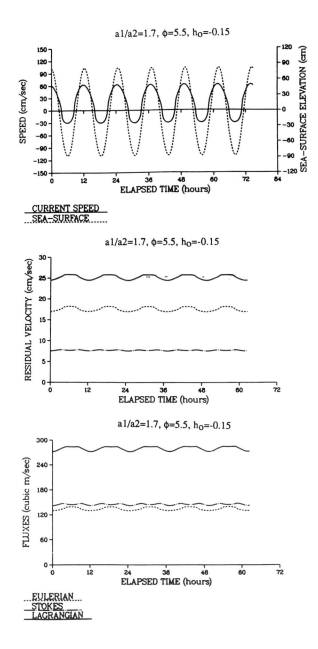

Figure 14. Theoretical simulation of instantaneous tidal currents, sea-surface, residual currents, and residual transports in West Channel.

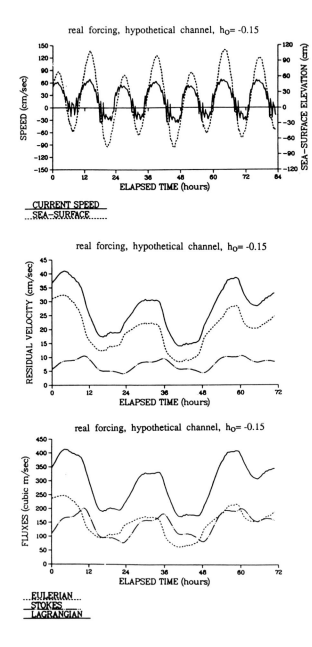

Figure 15. Simulation of instantaneous tidal currents, sea-surface, residual currents, and residual transports by applying the 'real-time' forcings and h_o = -0.15 m through the model channel.

SENSITIVITY OF RESIDUAL CURRENT–GENERATION
TO THE AMPLITUDE DIFFERENCE

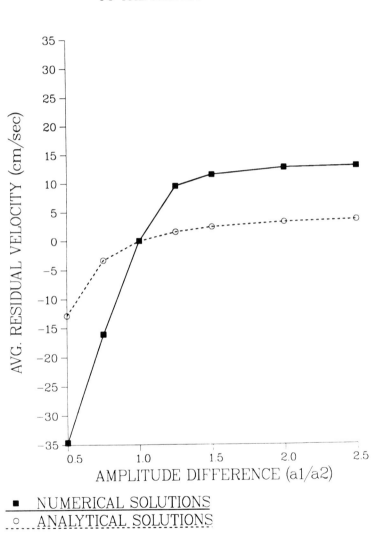

■ NUMERICAL SOLUTIONS
○ ANALYTICAL SOLUTIONS

Figure 16. Model comparisons in cases of:
(a) Tidal amplitude difference

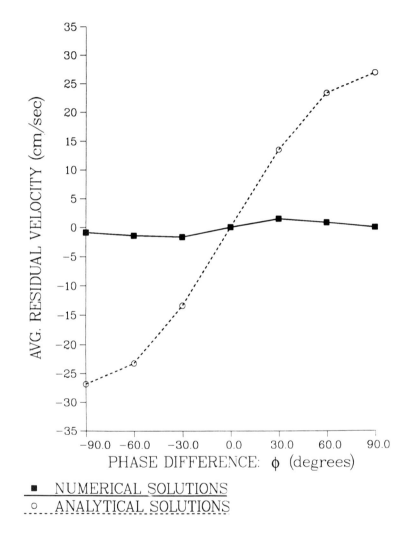

(b) Tidal phase difference

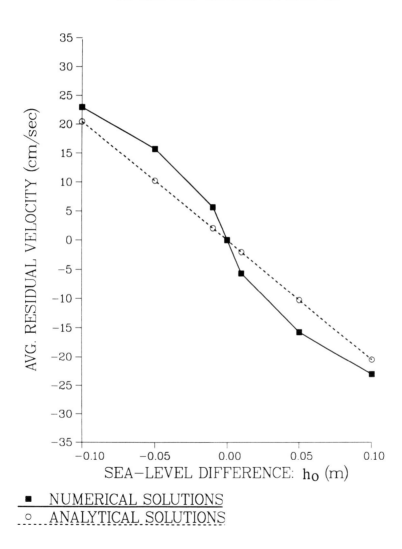

(c) Mean sea-level difference

level differences to the residual currents in the channel, respectively. Equation 9 also suggests that these contributions are linear. The two models agree well in the case of mean sea-level differences (Fig. 16c), they agree marginally in the case of amplitude differences (Fig. 16a), and they disagree significantly in the case of tidal phase differences (Fig. 16b). The disagreement reflects the inadequate consideration of non-linear effects by the analytical solution, which effects have been shown by our model to be important. For parameters that characterize the residual currents in West Channel, (a_1/a_2 = 1.7, ϕ = the tidal phase, amplitude, and mean sea-level differences are proportional to 1: 2 : 20. The analytical model concurs with the numerical model findings that the mean sea-level difference is the most influential factor in the generation of residual currents. However, the magnitude of the residual currents predicted by the analytical model is 35 cm/sec, compared to 18 cm/sec predicted by the numerical model. The higher value from the analytical model is probably due to the unrealistic linear expression of the three contributions in the solution. For instance, the numerical result of the Eulerian residual current for $a_1/a_2 = 1.7$, ϕ and $h_o = 0$, is 5 cm/sec; for $a_1/a_2 = 1$, $\phi = 0$, and $h_o = -0.1$, is 22 cm/sec. But for $a_1/a_2 = 1.7$, $\phi = 5.5°$, and $h_o = -0.15$, the residual current speed is only 18 cm/sec. Clearly the numerical results do not simply add up as the analytical solution suggests. Therefore, the influence of the tidal amplitude, phase, and mean sea-level differences on the generation of residual currents is interactive and nonlinear, and can only be expressed clearly by nonlinear numerical modeling.

Conclusions

a) The generation of residual currents in an open tidal channel is most sensitive to the mean sea-level differences, less sensitive to the tidal amplitude differences, and least sensitive to the tidal phase differences between the two ends of the channel.

b) The tidal phase difference between the two open boundaries of a channel is the most conducive to generating the M_4 overtide. The tidal amplitude differences have intermediate effect on the generation of M_4, and the mean sea-level differences have little effect on M_4 generation.

c) The field data depicting currents in West Channel can be fitted best by the model results for $a_1/a_2 = 1.7$ and a phase difference of $f = 5.5°$ for M_2 tide, and a mean sea-level difference of $h_o = -0.15$ m between south Chatham Harbor and Nantucket Sound. Sea-level in south Chatham Harbor is 15 cm higher than that in Nantucket Sound.

d) The residual current and M_4 overtide are generated by different mechanisms. The former is related more strongly to the friction within the system, and the latter is related more closely to the kinematic non-linearity in the system.

e) The combined influence of the tidal amplitude, phase, and mean sea-level differences between the two ends on the generation of residual currents in a tidal channel is interactive and non-linear, and can only be described properly by non-linear numerical models.

f) The historical evidence on the dominant westward sediment transport pattern observed in West Channel and the Monomoy breach suggests the existence of a mean current flowing from the Atlantic Ocean into Nantucket Sound. Field observation of tidal currents in West Channel indicates a strong westward residual current, having an average speed of 26 cm/sec. The tidal characteristics and residual flow in West Channel can be attributed best to the tidal amplitude and phase differences of the M_2 tides, but primarily a mean sea-level difference of 15 cm, between south Chatham Harbor and Nantucket Sound. Although the value of 15 cm may only apply to the 7-day period of the synoptic measurements of tidal heights and currents around West Channel, morphological data suggest that this difference may persist. Field data are inadequate to resolve this issue.

Acknowledgments

This study was partially funded by the Town of Chatham, the Commonwealth of Massachusetts through the Department of Environmental Management, the Coastal Engineering Research Center of the U.S. Army Corps of Engineers, the Coastal Research Center of Woods Hole Oceanographic Institution (WHOI), and NOAA National Sea Grant No. NA86-AA-D-SG090, WHOI

Sea Grant Project No. R/O-6. The U.S. Government is authorized to produce and distribute reprints for governmental purposes notwithstanding any copyright notation that may appear hereon. We thank Drs. Richard Signell and Wayne R. Geyer for reviewing an early version of the manuscript. Woods Hole Oceanographic Institution Contribution No. 8318.

References

Aubrey, D.G. and C.T. Friedrichs, 1988. Seasonal climatology of tidal non-linearities in a shallow estuary. In: *Hydrodynamics and Sediment Dynamics of Tidal Inlets,* Aubrey, D.G. and L.Weishar (eds.), Springer-Verlag, NY, p. 103-124.

Cotter, D.C., 1974. Tide-Induced Net Discharge in Lagoon-Inlet Systems. Master's thesis, University of Miami, 40 pp.

Giese, G.S., 1988. Cyclical behavior of the tidal inlet at Nauset Beach, Chatham, Massachusetts. In: *Hydrodynamics and Sediment Dynamics of Tidal Inlets,* Aubrey, D.G. and L. Weishar (eds.), Springer-Verlag, NY, p. 269-283.

Heath, R.A., 1980. Phase relations between the over- and fundamental-tides. *Dt. hydrogr. Z.,* v. 33 H (5), p. 177-191.

Hine, A.C., 1975. Bedform distribution and migration patterns on tidal deltas in the Chatham Harbor Estuary. In: *Estuarine Research*, v. 2, Cronin, L.E. (ed.), Academic Press, Inc., NY, p. 235-307.

Huang, P.-S., D.-P. Wang, and T.O. Najarian, 1986. Analysis of residual currents using a two-dimensional model. In: *Physics of Shallow Estuaries and Bays,* van de Kreeke (ed.), Springer-Verlag, NY, p. 71-80.

Ianniello, J.P., 1977. Tidally induced residual currents in estuaries of constant breadth and depth. *J. Mar. Res.,* v. 35, p. 755-786.

Ianniello, J.P., 1979. Tidally induced residual currents in estuaries of variable breadth and depth. *J. Phys. Ocean.,* v. 9, p. 962-974.

Liu, J.T., D.K. Stauble, G.S. Giese, and D.G. Aubrey, this volume. Morphodynamic evolution of a newly formed tidal inlet. In: Aubrey, D.G. and G.S. Giese, (eds.), *Formation and Evolution of Multiple Inlet Systems, Coastal and Estuarine Studies Series,* AGU.

Pingree, R.D. and L. Maddock, 1977. Tidal residuals in the English Channel. *J. Mar. Biol. Ass. U.K.,* v. 57 (2), p. 339-354.

Prandle, D., 1978. Residual flows and elevations in the southern North Sea. *Proc. R. Soc. Lond. A.,* v. 359, p. 189-228.

Speer, P.E., and D.G. Aubrey, 1985. A study of non-linear tidal propagation in a shallow estuarine system, part 2: theory. *Estuarine, Coastal and Shelf Sci.,* v. 21, p. 207-224.

Tee, K.T., 1976. Tide-induced residual current, a 2-D non-linear numerical model. *J. Mar. Res.,* v. 34, p. 603-628.

Tee, K.T., 1977. Tide-induced residual current-verification of a numerical model. *J. Phys. Oceanogr.,* v. 7, p. 396-402.

van de Kreeke, J., 1978. Mass transport in a coastal channel, Marco River, Florida. *Estuarine, and Coastal Mar. Sci.,* v. 7, p. 203-214.

van de Kreeke, J., 1980. Tide-induced residual flow. In: *Mathematical Modeling of Estuarine Physics,* Sundermann, J. and K.-P. Holz, (eds.), Springer-Verlag, NY, p.133-144.

van de Kreeke, J. and R.G. Dean, 1975. Tide-induced mass transport in lagoons. *J. Waterways, Harbor and Coastal Eng. Div.,* v, 4, p. 393-403.

Wong, K.-C., 1989. Tidally generated residual currents in a sea-level canal or tidal strait with constant breadth and depth. *J. Geophys. Res.,* v. 94, no. 6, p. 8179-8192.

5

Backbarrier and Inlet Sediment Response to the Breaching of Nauset Spit and Formation of New Inlet, Cape Cod, Massachusetts

Duncan M. FitzGerald and Todd M. Montello

Abstract

The breaching of Nauset Spit formed an inlet 2 km in width, increasing the tidal range (0.3 m) and tidal currents in the backbarrier. Wave action and flood tidal currents delivered a large quantity of sand to the harbor region which was derived from the beach and littoral system. The influx of sand and stronger tidal exchange between the ocean and bay have produced changes in backbarrier channel morphology and position and extent of the intertidal sand bodies and subtidal shoals. Between January 1987 and August 1990 the flood-tidal delta migrated northward an average of 150 m along its northern boundary and its intertidal area doubled in size. This landward movement of sand is countered by seaward transport of sediment in the ebb-dominated channels adjacent to the delta. It is presumed that increased tidal current strength in East Channel was responsible for erosion of a large, ebb-oriented, spillover lobe that existed at the south end of the delta prior to the breaching.

The sediment of Chatham Harbor is composed of medium-coarse, moderately well-sorted sand. Some gravel is found in the deeper sections of the channels. The channels and sand bodies are covered with a variety of complex and simple bedforms; however, simple crested sandwaves dominate. Sandwaves in the channels have spacings ranging from 16 to 42 m with heights of

Formation and Evolution of Multiple Tidal Inlets
Coastal and Estuarine Studies, Volume 44, Pages 158-185

0.5 to 3 m while those of the sand bodies and shoals are lower (< 1 m) with a greater range in spacing (8 to 69 m). Bedform orientations and current measurements have been used to estimate net sand transport trends. The back-barrier is dominated by several sand circulation gyres.

Introduction

The infrequency of barrier breaching during storms explains why little is known concerning how inlets reach dynamic equilibrium. Due to its accessibility, the breaching of Nauset Spit during the northeast storm of 2 January 1987 provided an excellent opportunity for scientists to document the evolution of a tidal inlet and its backbarrier environment (Fig. 1). The growth of New Inlet to 1 km in width after just five months of its formation generated nation-wide interest as the ensuing wave erosion that occurred along the Chatham shoreline landward of the inlet destroyed several houses and endangered millions of dollars worth of property.

The breaching of Nauset Spit has produced substantial changes in tidal hydraulics, erosional-depositional trends and channel and shoal morphology in Chatham Harbor. One of the immediate effects was an increase in tidal range within the harbor of 0. 2 to 0. 3 m (Liu et al., this volume). Some of the sand that had once comprised Nauset Spit and other sediment that had been part of the southerly longshore transport system was moved into Chatham Harbor, building sand shoals and bedforms. In addition, some of the sand went into the construction of the inlet's large ebb-tidal delta shoal (Fig. 2).

The purpose of this investigation was to determine the response of the backbarrier to the breaching of Nauset Spit, encompassing the region between Tern Island and Nauset Spit southward to the inlet mouth. The study focused on delineating the morphological changes of the channels, shoals and bedforms as well as determining the net sediment transport patterns.

Background

Nauset Spit is part of a barrier spit-barrier island complex that extends

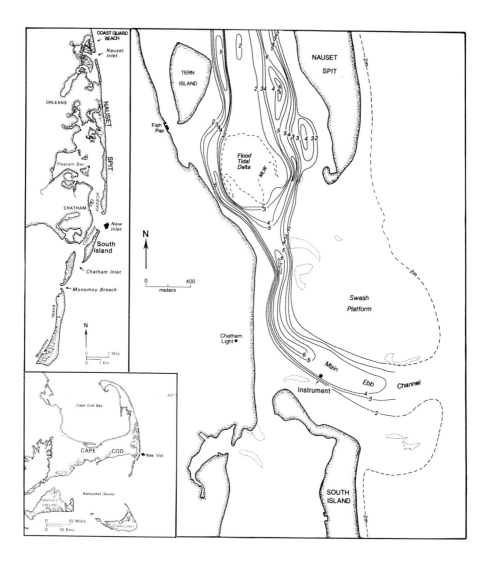

Figure 1. Location of study area.

southward from Coast Guard Beach to Monomoy Island with several inter-
vening tidal inlets. The barriers, which built southward through spit accretion
from sediment derived from eroding glacial cliffs, have served to straighten
the irregular, late Holocene shoreline forming a semi-continuous system of
bays and lagoons with peripheral tidal marshes.

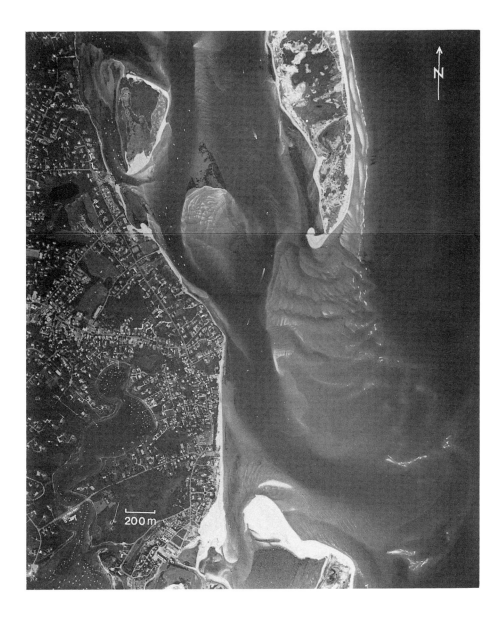

Figure 2. August 1990 vertical aerial photograph of New Inlet.

The breaching of Nauset Spit, which has occurred many times during the past 220 years (McClennen, 1979; Giese, 1988) and which had been predicted by Giese (1988), is related to a gradual restriction of tidal flow through Chatham Inlet. This produces differences in tidal range and tidal phase between the ocean and the upper bay which causes a substantial hydraulic head across the barrier (Aubrey et al., this volume). The long-term erosion and thinning of Nauset Spit (Gatto, 1978; Cornillon, 1979) coupled with a storm surge of 0.5 to 1.0 m and waves over 3.0 m high (Giese, 1990) associated with the 2 January 1987 northeaster facilitated overwashing of the barrier and eventual channel formation by ebb tidal currents. Initially, the overwash channel was shallow and only 6 m wide; however, tidal exchange quickly deepened and widened the inlet to 520 m by March 1987 and 1,715 m (throat section) by August 1990 (Liu et al., this volume).

New Inlet has a mean tidal range of 1.75 m which increases to 2.3 m (at the inlet throat) during spring tide conditions. Most of the tidal prism of New Inlet comes from the exchange of tidal waters between the ocean and Chatham Harbor and Pleasant Bay to the north. Presently, the inlet spit associated with South Island has almost closed off the entrance to the southern portion of Chatham Harbor and, therefore, most of its tidal waters exit and fill through inlets to the south. From the inlet throat the main channel turns northward and bifurcates around a large flood tidal delta (Fig. 2). North of the delta the East and West Channels are separated by a north-south trending subtidal shoal. Maximum channel depth in the backbarrier is 7 m. Previous studies of Chatham Harbor investigated the distribution of large scale sand bodies and the distribution of bedforms on these bars (Hine, 1972; 1975). Unfortunately, this prior study did not provide much detailed information for the northern portion of the harbor where the present project is located.

Methods

Morphological changes to the Chatham Harbor region resulting from the breaching of Nauset Spit as well as the sedimentation processes that produced these changes were determined from several field and laboratory studies undertaken during the late summer and fall of 1990. Additional information concerning the study area was obtained from historical records and sequential vertical aerial photography.

Continuous depth recordings of the major channels and subtidal shoals in Chatham Harbor were taken with a DE-719B Raytheon fathometer having a resolution of +/- 10 cm. Longitudinal profiles of the channels were made to map the distribution and orientation of bedforms and major subtidal shoals. The morphology of the channels was determined from the longitudinal transects and 12 cross-sectional profiles. Navigation for the profiles was through dead-reckoning using numerous landmarks and buoys that were discernible in large scale aerial photographs of the region. Bedform terminology in this paper will follow that of Boothroyd and Hubbard (1975).

The relatively small size of the flood-tidal delta allowed it to be mapped using a measuring tape and Brunton compass. Bedform orientations, heights, and wave lengths were noted at 50 m intervals throughout the delta. Numerous stakes were positioned at bedform crests and at the edge of the ebb shield to monitor migrational patterns during a period of approximately two weeks. Longer-term changes in the study area were determined from vertical aerial photographs obtained from Col-East Inc. in North Adams, MA. One set of photographs of the region, taken in 1982, provided a baseline for comparing changes in the backbarrier after the inlet was opened. Eight subsequent photoflights were rode of the study area from 7 May 1987 to 15 August 1990 with flights spaced every 4 to 8 months (Table 1). A summary of the geomorphic changes in the backbarrier is given in Table 2.

Sediment distributions in the backbarrier were determined from 28 surface (8-20 on deep) samples collected on the flood-tidal delta (11), swash platform (5), and major channels (12). A Van Veen grab sampler was used to obtain samples in the deep channels. Position of the samples on the flood-tidal delta was recorded during the mapping project whereas the rest of the sampling sites were located using navigational aids and aerial photographs. Detailed enlargements of color aerial photographs and the presence of easily recognizable landforms enabled a close approximation (+/- 10 m) of the sampling sites in the field. Size fraction analyses were performed on the samples using standard techniques and employing nested sieves.

Table # 1 Vertical Aerial Photographs

Month/Year	Set #	Date of Photograph	Print	Scale
	1	20 October 1982	Color	1 : 9000
Jan./1987		[Inlet opened 2 January 1987]		
Apr.				
	2	7 May 1987	Black & White	1 : 9000
Jul.				
Oct.				
Jan./1988	3	11 January 1988	Black & White	1 : 9000
Apr.				
	4	5 May 1988	Black & White	1 : 9000
Jul.				
Oct.	5	1 September 1988	Black & White	1 : 9000
Jan./1989	6	29 December 1988	Black & White	1 : 9000
Apr.				
	7	23 May 1989	Black & White	1 : 9000
Jul.				
Oct.				
Jan./1990	8	7 December 1989	Black & White	1 : 9000
Apr.				
Jul.				
	9	15 August 1990	Color	1 : 9000
Oct.				

TABLE 2 Historical Morphological Changes Of The Backbarrier In The Vicinity Of The Flood Tidal Delta

Date	West Channel	Flood Tidal Delta	Mussel Bed	East Channel	East Channel Bar
20 October 1982	Dominated by ebb oriented sandwaves.	Well developed horse-shoe shape. Dominated by flood oriented sandwaves.	Beds abut the flood tidal delta/ (FTD) on the northern, eastern, and southern sides.	Ebb and flood channels present. Ebb channel occupies the eastern portion of the channel and the flood channel occupies the western portion. An ebb oriented sand lobe is located at the mouth of the ebb channel.	Dominated by flood oriented sandwaves.
7 May 1987	Dominated by ebb oriented sandwaves. An ebb oriented sand lobe is beginning to form at the mouth of the channel.	Distorted horse-shoe shape. Beginning to migrate north.	Beds abut the FTD to the north only.	A flood oriented inter-tidal sand bar is migrating northward into the mouth of the channel.	Large ebb oriented sandwaves now occupy the east flank of this bar. Smaller flood oriented sandwaves occupy the seaward side of this bar.
11 January 1988	Dominated by ebb oriented sandwaves. The ebb oriented sand lobe still develops at the mouth of the channel.	An intertidal bar begins to migrate onto the southeastern portion of the FTD. An ebb spillover lobe forms on the west side of the FTD.	Beds abut the FTD to the north and to the east.	An intertidal bar is migrating north up the channel. The flood channel is migrating seaward and deteriorating.	The bar is becoming segmented due to the development of an ebb spillover lobe. The bar is still dominated by ebb oriented sandwaves.
5 May 1988	Dominated by ebb oriented sandwaves. The ebb oriented sand lobe at the mouth of the channel is now well developed. The channel is constricted near the mouth due to the development of the ebb spillover lobe on the west side of the FTD.	Ebb spillover lobe is still evident on west side of the FTD An intertidal sand bar is still migrating to the north across the FTD. A complex pattern of bedforms is now evident on the FTD. This pattern indicates flow directions over the FTD to be to the northeast and northwest.	Beds abut the FTD to the north and east.	Flood oriented lobe still evident in the channel. Sediment from this lobe is being reworked onto the FTD.	Segmented by a well developed ebb spillover lobe. Ebb oriented sandwaves dominate this bar. A flood oriented sand lobe is present at the north end of the flood channel that is on the east flank of this bar. This channel is migrating seaward.
1 September 1988	Dominated by ebb oriented sandwaves. The channel is constricted at the mouth due to continued ebb spillover lobe development on the west side of the FTD. A portion of the sand lobe at the mouth of the channel has become distended seaward.	Ebb spillover lobe is still evident on the west side of the FTD. An ebb spit is forming on the east side of the FTD and extends diagonally seaward across the East Channel and joins the swash platform at the southern end of the East Channel Bar.	Beds abut the FTD to the north.	No change since May 1988.	The bar is segmented by an ebb spillover lobe. The bar is dominated by ebb oriented sandwaves. The flood channel is deteriorating.
29 December 1988	Dominated by ebb oriented sandwaves. The sand lobe at the mouth of the channel is deteriorating. An ebb spit projects out into the channel near the mouth from a mainland promentory.	The FTD is beginning to take on the classic trilobite shape. A well developed ebb spit exists on the east side of the FTD. One begins to form on the west side.	Beds abut the FTD to the north.	No change since May 1988.	The bar is migrating seaward. Ebb oriented sandwaves dominate the landward side of the bar. Flood oriented sandwaves dominate the seaward side of the bar.
23 May 1989	Dominated by ebb oriented sandwaves. Ebb spit continues to develop off of the mainland promentory and constrict the channel at the mouth. The sand lobe at the mouth has deteriorated.	The FTD obtains primitive trilobite shape. Multiple ebb spits are evident on the East side of the FTD. The ebb spit on the west side continues to develop. The inter-tidal sand bar continues to migrate over the FTD.	Beds abut the FTD to the north.	Ebb oriented bedforms are becoming more evident. Flood oriented bedforms still exist in the eastern portion of the channel.	The bar continues to migrate seaward.
7 December 1989	Dominated by ebb oriented sandwaves. A sand lobe is again evident at the mouth of the channel.	The FTD maintains the primitive trilobite shape. The intertidal sand bar is more extensive and still migrating north across the FTD.	Beds abut the FTD to the north.	The channel boundaries are farther defined near the mouth. Ebb flow is constricted to the east by the FTD (ebb spit). To the west, the swash platform constricts flow.	Dominated by ebb oriented sandwaves. The flood channel is well developed on the seaward side of the bar.
15 August 1990	Dominated by ebb oriented sandwaves. Channel is constricted by an ebb spit that is developing off of a mainland promentory. This ebb spit projects out into the channel. The sand lobe at the mouth has once again deteriorated.	The FTD obtains the classic trilobite shape. Well developed ebb spits flank the FTD on both the east and west sides.	Beds abut the FTD to the north.	The channel is still constricted at the mouth by the flood ramp to the west and by the swash platform to the east.	Flood channel that existed on the seaward side of the bar is gone. Ebb oriented sandwaves dominate the bar.

A cursory examination was made of the relative magnitude and flood or ebb dominance of tide-generated currents at four stations in Chatham Harbor and at the inlet threat. Current velocities and directions were made using a Marsh McBirney model 201D digital electromagnetic current meter having an accuracy of 5-10%. At each station readings were depth-averaged and recorded each hour over an approximate 13 hr tidal cycle. In addition to these measurements an instrument package (Sea Data 635-12) consisting of a quartz pressure sensor, temperature probe and X-Y coordinate electromagnetic current meter was installed at the inlet throat in 5 m of water with the current probe mounted 1 m off the bed (Fig. 1). The gauge was deployed for a 30-day period from 18 September to 18 October 1990 and during this time current velocity and direction and tidal elevation were sampled 32 times every hour. Although a detailed discussion of these data is outside the scope of this paper, these measurements were helpful in delineating the current dominance of the inlet throat and in evaluating the velocity data gathered at the stations in the harbor region.

Flood-Tidal Delta

In Chatham Harbor there is one major flood-tidal delta (Fig. 2) having a classic trilobate shape, morphological features, and sedimentary processes as first described by Hayes (1975). The delta is located at the widest section of the Harbor and is bordered on the north by an extensive mussel (*Mytilus edulis*) community. The flood delta has been present in some form for at least the past 30 years and probably much longer. Hine (1972) describes it as an S-shaped sand body (cf., Caston, 1972) with the different lobes oriented into the dominant flood and ebb currents. He theorizes that the delta formed during a time when a tidal inlet was located at this position from sand that was transported into the inlet by flood currents. As the tidal inlet and updrift spit system migrated to the south, the sand body was reworked by tidal currents and became realigned by the north-south flow. Hine (1972) identified the sand bodies at Chatham Light and Horne's Marina as having similar histories.

A map constructed of the delta on 28 September 1990 (Fig. 3) indicates that the intertidal portion of the shoal is dominated by flood-oriented sandwaves with spacings varying from 8 to 31 m and heights generally less than 0.7 m

used to construct a bathymetric map of the study area (Fig. 1). Figures 1 and 6 indicate that the main ebb channel gradually deepens from approximately 2.0 m at the terminal lobe to 3.2 m between the channel margin linear bars (Section G-G', Fig. 6) and 5.5 m at the inlet throat (Section F-F', Fig. 6). From the inlet throat the channel bends to the north and becomes confined between the swash platform and the mainland shoreline. Channel scour in this region

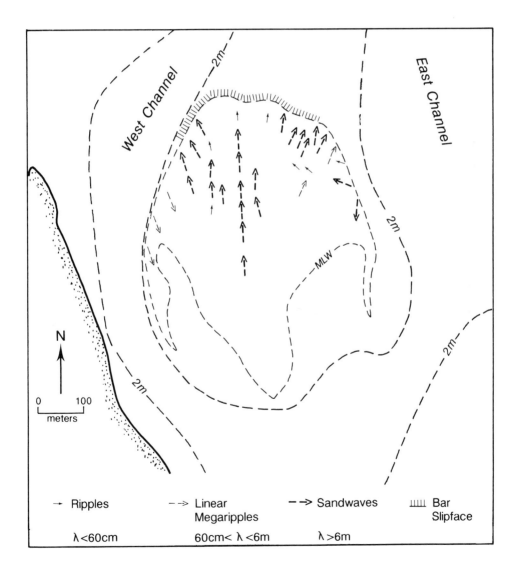

Figure 3. Map of bedform distributions and orientations on the flood-tidal delta on 28 September 1990.

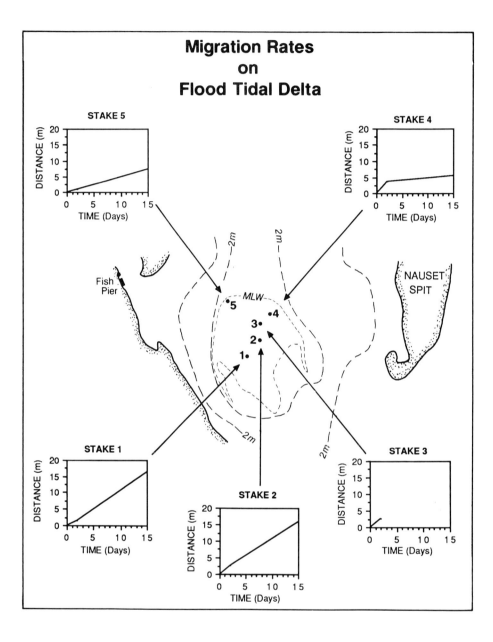

Figure 4. Northerly migration rates of sandwaves (#1-4) on the flood-tidal delta and migration rate of ebb shield portion of the flood delta (#5) determined between 28 September and 12 October 1990.

and averaging 0.5 m. Ebb-oriented linear megaripples dominate the ebb spits. Ripples occur throughout the delta and commonly indicate secondary transport patterns in the troughs of the sandwaves.

Monitoring of stakes placed at the crest of several sand waves demonstrated that some bedforms are migrating at rates of 0.5 to 1.0 m/day (Fig. 4). The lower migration rate at Station #3 is best explained by the fact that a complete bedform wavelength had probably migrated past the stake and the second measurement was not made to the original sand wave crest. The spacing of the sandwave at Station #3 was 20 m. The lack of longer-term data at Station #4 is due to the uncertainty as to which bedform crest to measure. Here the average bedform wavelength is only 8 m and the original sandwave crest could not be identified during the second reading period (day 15). Station #5 (Fig. 4) was positioned to monitor the stability of the ebb shield. During a 15-day period the slipface that defines the northern extent of the delta had migrated northward 7.5 m or 0.5 m/day. Overlays of the intertidal boundary of the flood-tidal delta between 1982 and 1990 revealed that the delta had enlarged substantially and has been migrating to the northeast (Fig. 5, Table 2). If it is assumed that the opening of New Inlet and the ensuing increased tidal flow initiated these changes and that the delta was stationary prior to this time, then since January 1987 the ebb shield has migrated between 130 and 210 m. At Station #5 the northward migration of the slipface was 150 m, an average rate of 0.2 m/day. The much greater migration rate (3x) measured more recently may suggest that the ebb shield undergoes periodic erosion, perhaps during storms.

The overlays also demonstrate that since the breaching the flood delta has enlarged significantly (Fig. 5), doubling in areal extent to 2.0×10^5 m^2 by May 1990 (Liu et al., this volume). During the past three years the western margin of the delta has accreted westward and eroded back to the east. Presently, the margin is slightly eastward of its position in 1982.

A secondary, mostly subtidal shoal exists immediately landward of the inlet between Lighthouse Beach and the inlet spit system of South Island (Fig. 2). This location had been the site of a flood-tidal delta prior to the formation of New Inlet (Hine, 1972; Liu et al., this vol.), but since that time the sand deposit has been reworked by wave and tidal processes (Weidman and Ebert, this

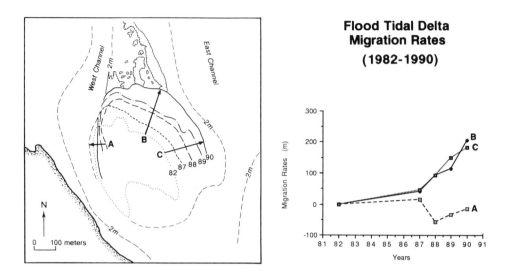

Figure 5. Northeasterly migration of the flood-tidal delta from 1982 to August 1990 as determined from vertical aerial photographs.

volume) and lacks the shape and features associated with a flood delta (cf., Hayes, 1975). For example, the sand waves which presently cover the shoal are ebb- rather than flood-oriented (Fig. 2). The shoal is influenced predominantly by ebb currents which issue from the bay behind South Island and ebb flow within the main inlet channel. Weidman and Ebert (this vol.) have shown that this portion of the backbarrier is constantly changing and dominated by a cyclic process in which the inlet spit system undergoes accretion, breaching and finally shoal generation. The reader is directed to their paper for a more detailed discussion of this area.

Backbarrier Channels

Morphology

Fathometer cross-sectional profiles illustrate the morphology of backbarrier channels (Fig. 6). These profiles together with the longitudinal transects were used to construct a bathymetric map of the study area (Fig. 1). Figures 1 and 6 indicate that the main ebb channel gradually deepens from approximately 2.0 m at the terminal lobe to 3.2 m between the channel margin linear bars (Section G-G', Fig. 6) and 5.5 m at the inlet throat (Section F-F', Fig. 6). From the inlet throat the channel bends to the north and becomes confined between the swash platform and the mainland shoreline. Channel scour in this region produces the greatest depth in the harbor (7.0 m) just south of Section E-E' (Fig. 6). The ebb spit which extends southeastward from the mainland shoreline (Fig. 2) also is clearly visible in Section E-E' (Fig. 6). The effects of the bifurcating tidal flow around the flood delta are seen in Section D-D' (Fig. 6) from the incipient development of partitioned channel areas. Sections B-B' and C-C' (Fig. 6) indicate that the channels on either side of the flood delta are between 4.3 and 5.0 m deep. North of the delta, East and West Channels are separated by a wide, sandwave-covered shoal less than 2.0 m deep (Section A-A', Fig. 6).

The sequential aerial photographs of the backbarrier indicate that East Channel was most affected by the breaching event (Table 2). It appears that the increased tidal flow resulting from the formation of New Inlet enlarged the channel by removing the large ebb-oriented, spillover lobe that was present at the confluence of the East and West Channels in the 1982 photograph and described by Hine (1972). However, the channel and adjacent shoal complex that abuts Nauset Spit still retain areas of flood and ebb dominance as indicated by bedform orientations and channel shoaling patterns (Table 2). The morphological changes documented by the sequential photographs demonstrate that the backbarrier has not yet fully evolved to a state of dynamic equilibrium and further changes can be expected until the inlet migrates south and tidal and wave energies are gradually reduced.

Bedform Patterns

Due to the strong tide and wave-generated currents acting on the sandy backbarrier substrate, most of the channels and shoals in the study area are covered with bedforms ranging in scale from ripples and megaripples to sand-waves (Fig. 7). In addition to the fathometer profiles, aerial photographs and field measurements have been used to produce a bedform distribution map of the backbarrier and inlet areas (Fig. 8). The most common bedforms in the study area are sandwaves which are usually asymmetric (e.g., Section B-B'

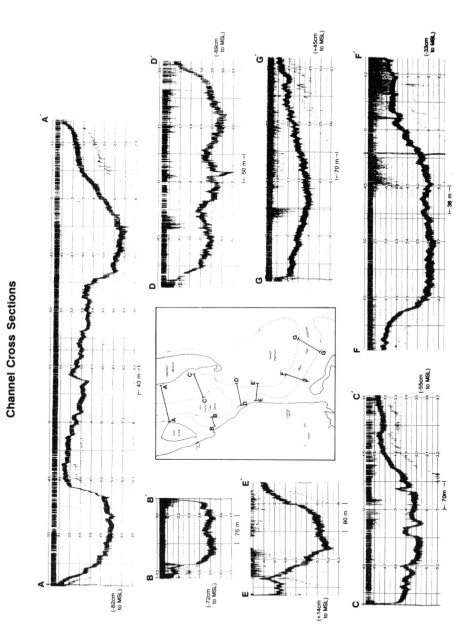

Figure 6. Channel cross sections throughout Chatham Harbor and New Inlet determined from fathometer profiles.

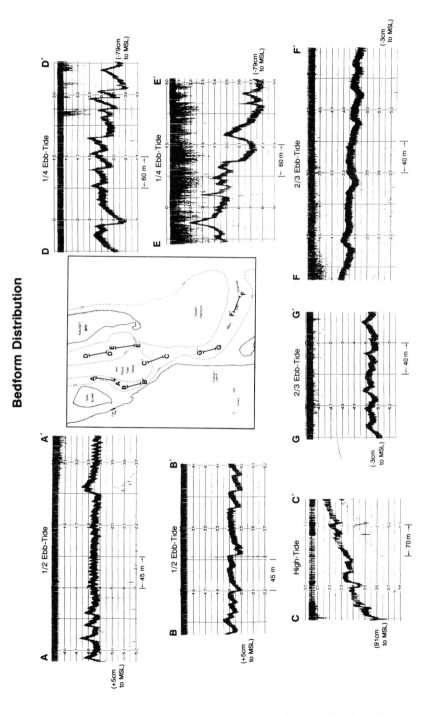

Figure 7. Fathometer profiles illustrating bedform distributions in the major channels.

and C-C', Fig. 7), however, symmetrical forms (e.g., Section E-E' and G-G', Fig. 7) are present as well. Megaripples are found commonly on the backs of many sandwaves (e.g., Section D-D', Fig. 7) and occur in fields by themselves (e.g., Section A-A', Fig. 7).

As seen in Figures 7 and 8, the bedform data suggest that flow in the backbarrier is segregated. The flood-tidal delta is dominated by floodoriented sandwaves having spacings varying from 8 to 63 m with heights less than 1 m. The linear subtidal shoal that extends northward from the delta as well as the adjacent West Channel and shallow portion of the East Channel are covered by flood-oriented, moderately-spaced (23-25 m) sandwaves. In the adjacent deeper section of the East Channel ebb- and flood-oriented bedforms are directed toward a zone of aggradation where channel depths shoal to less than 3 m (Fig. 7). South of this region in East Channel and adjacent to the northern end of the flood-tidal delta in West Channel, there are locations where bedforms migrate in opposite directions (Fig. 8). The nodal points of these two systems mark the beginning of the ebb dominance of the main inlet channel, a trend which extends to the terminal lobe of the ebb-tidal delta. Large southeasterly-oriented transverse bars exist along the southern distal portion of the ebb-tidal delta (Figs. 2 and 8).

Little is known concerning the bedform distributions and bar migrational trends of the swash platform, other than what can be discerned from the vertical aerial photographs. The historical photographs indicate that this shallow, mostly subtidal feature has undergone complicated changes during the widening of the inlet and development of the platform (Liu et al., this volume). Presently, the area is arranged into crescentic bar regions separated by shallow channels. Some of the bars appear to be flood-dominated whereas others seemed to be migrating seaward. Considerably more fieldwork is needed on the swash platform to document this portion of the inlet sediment transport system.

Grain Size Distributions

The entire study area is dominated by medium-to-coarse sand, while some gravel exists in portions of the major channels. It seems likely that most of the

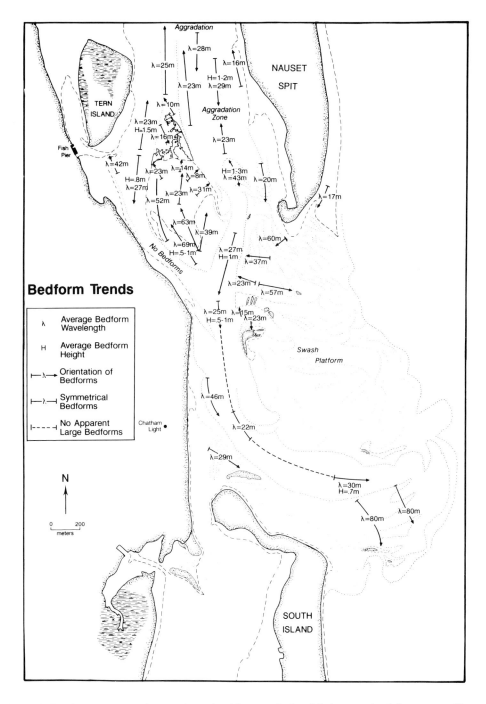

Figure 8. Bedform summary diagram determined from vertical aerial photographs, fathometer profiles, and field measurements.

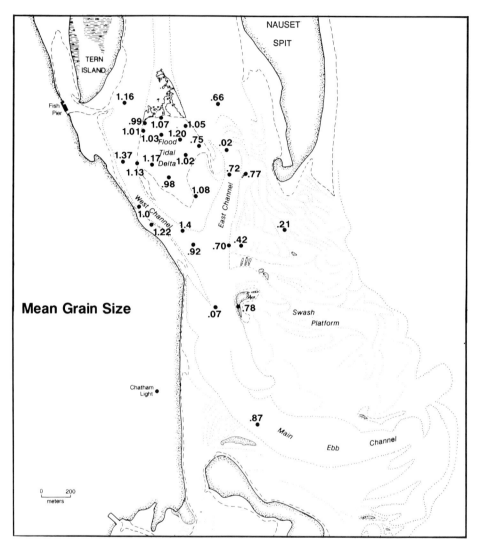

Figure 9. Mean grain sizes (units in phi).

sediment in the backbarrier was derived from the breaching and erosion of Nauset Spit during the formation of New Inlet; this sediment was added to the sand which earlier had been transported landward through former tidal inlets along this coast. Secondary sources of sediment are erosion of the mainland shoreline and possibly channel scour of the underlying substrate. The original source of much of the sediment lies in the eroding glacial cliffs (Eastham Outwash Plain) north of Nauset Inlet. This material is transported to the inlet by the southerly longshore transport system.

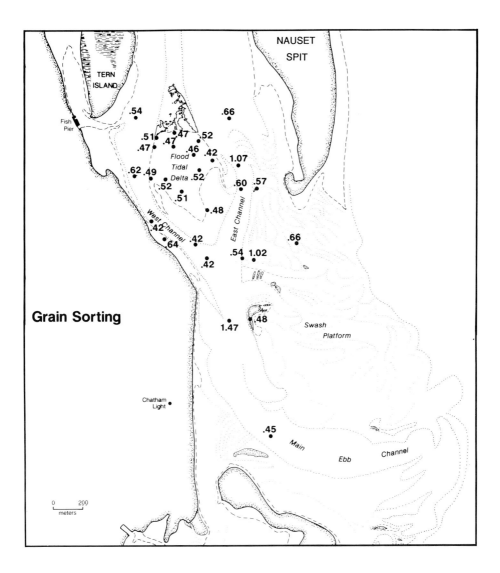

Figure 10. Grain sorting (traits in phi).

The 28 sediment samples indicate several interesting grain size patterns including flood-tidal delta sands which are liner-grained (medium sand) and slightly better sorted than the swash platform sands (coarse sand) (Figs. 9 and 10). Another interesting trend is that the main inlet channel and East Channel are coarser-grained than West Channel. The reason for this is unknown, but it may be due to the proximity of West Channel to the swash platform across which coarse sand movs westward and enters the backbarrier.

Two samples were much coarser than the other 26 samples: #13, (0.07 phi) located between the swash platform and the harbor shoreline, and #25 (0.02 phi), located in the East Channel between the flood delta and southern end of Nauset Spit. At both sites the sediment was bimodal containing a large percentage of gravel (> 90% by weight). The mean gravel size at Station #13 was 2.3 cm (long axis) including several clay clasts. These consisted of semi-indurated silty clay having a bluish gray color, perhaps of glacial lake origin (Oldale, pers. comm.). The grain size statistics given in Figures 9 and 10 only considered the sand fraction of the grab sample. The sand at the two gravel locations was also poorly sorted, whereas the rest of the samples were mostly well-sorted to moderately well-sorted (Fig. 10). Analyses of the grain size data indicate that more samples are needed to characterize definitively the sedimentary environments and document sediment transport trends.

Current Data

The current velocity time series recorded at various locations in the major channels on 29 September 1990 provide an indication of the magnitude of tidal currents at the study area and a basis for comparing the relative current strength among the five stations (Fig. 11). The measurements were made during neap tide conditions with an inlet throat ebb tidal range of 1.1 and flood range of 1.5 m (FitzGerald and Montello, 1990). The single data set indicates that the strongest ebb currents occur, as one would expect, at the inlet throat, having a maximum velocity of 83 cm/s. Even though the succeeding flood tidal range was 0.4 m greater than the ebb, the maximum flood current reached only 56 cm/s, 27 cm/s less than the ebb.

The strongest flood tidal currents were recorded at Stations 2 & 4 (Fig. 11) with velocities of 82 and 80 cm/s, respectively. At these same locations the preceeding ebb currents were weaker, particularly at Station #4 where maximum ebb currents were only 46 cm/s. At Station #2 the preceeding maximum ebb current was 68 cm/s. The apparent dominance of the flood currents at Station #2 is likely attributable to the 27% disparity in tidal range whereas at Station #4 the weaker ebb currents also may be a product of positioning problems. At Station #5 in the West Channel the flood currents were slightly weaker than in the East Channel whereas the opposite was true for the ebb currents. Station #3, which was located on the flood ramp of the delta, was clearly flood-dominated due to shielding from the ebb currents by the intertidal portion of the landward shoal.

The strong ebb dominance of the tidal currents at the inlet throat was corroborated by the long-term measurements recorded during the Sea Data 635-12 instrument deployment (FitzGerald and Montello, 1990) and in earlier studies by Liu et al. (this vol.). The ebb dominance is a product of the dynamic equilibrium that has been established among tidal exchange between the ocean and bay, sediment transport patterns, inlet migrational trends, and the channel morphology of the backbarrier. Although the opening of the inlet is quite wide (approx. 2,000 m at mouth), the channel cross section is highly asymmetric and most of its width (approx. 1,400 m) is comprised of a shallow swash platform with a narrow channel thalweg (Fig. 1). Because maximum flood currents occur close to high water, a considerable volume of the flood tidal prism enters the inlet over the intertidal platform area. Conversely, the maximum ebb currents in the inlet occur near low water when the swash platform has its greatest exposure. This condition coupled with the north-south elongation of the backbarrier, which produces a general southerly ebb flow confined to the main channel, results in a strongly ebb-dominated main inlet channel and a swash platform dominated by landward-directed currents (FitzGerald and Montello, 1990).

Net Sediment Transport Trends

Sediment transport patterns at Chatham Harbor and at the entrance to New Inlet have been determined from bedform and bar orientations, shoal migrational trends, current data, sediment analyses and historical information (Fig. 12). Sand enters the backbarrier primarily over the swash platform through a complicated pattern of migrating shoals and bedforms and through bedload transport in channel systems. The processes and pathways whereby this is accomplished are still unclear. It is assumed that most sand movement into Chatham Harbor occurs during storms when the storm surge and shoaling and breaking waves create stronger landward flow and greater sand suspension across the swash platform.

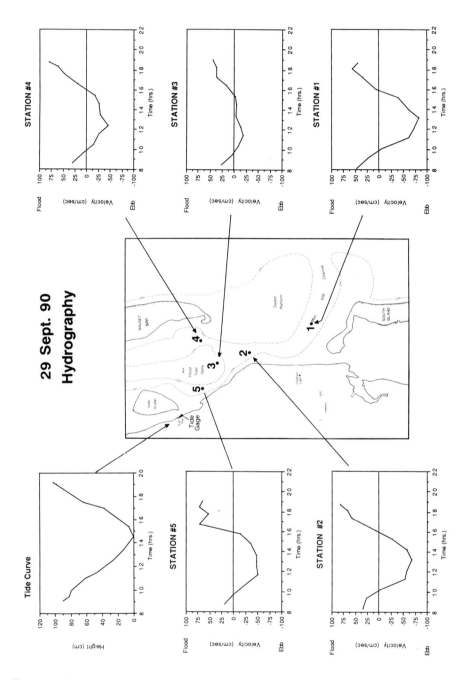

Figure 11. Velocity time series for five stations at the inlet throat and backbarrier region. The flood tidal range was about 40 cm greater than the ebb range during the measurement period. The tide curve was produced from measurements taken from a tide staff located at Fish Pier.

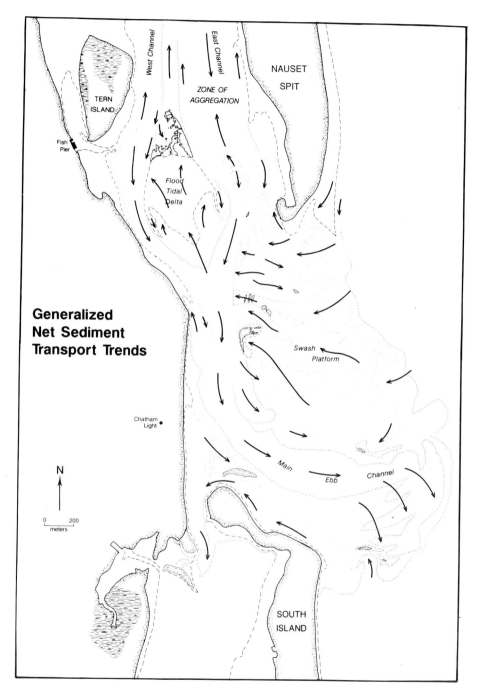

Figure 12. Summer, 1990, sediment transport patterns determined from current measurements, bedform and shoal slipface orientations and migrational trends, and grain size data.

Some sand that enters the inlet across the northern portion of the swash platform is circulated within the backbarrier. The prevalence of bedforms and recent changes in the morphology and position of shoals and channels, including the flood delta and East Channel, attests to the activity of sand movement within Chatham Harbor. Short- to long-term sediment sinks within the study area include the flood delta, convergent areas within East Channel, the longitudinal shoal between West and East Channels, and perhaps regions in Pleasant Bay. Two major circulation cells within the harbor involve the northward movement of sand from the channel onto the flood-tidal delta at the confluence of East and West Channels. Due to the growth and migration of the delta, some of this sand may be trapped within the delta for a long period of time (> 10 yrs) before it is exhumed and becomes part of the active transport system again. Eventually, sand transported across the delta is dumped into the East and West Channels where it is moved in a net southerly direction by the dominant ebb tidal currents (Fig. 11). Some of this sand may be moved back onto the flood delta completing the sediment gyre.

South of the confluence of East and West Channels sand in the main inlet channel is transported in a net seaward direction to the distal portion of the ebb-tidal delta. The large southeasterly-trending transverse bars that occupy this region (Fig. 2) are part of the inlet sediment bypassing system. Some of the sand that moves onshore from the southern, distal portion of the ebb delta is reintroduced into the inlet through the southern marginal flood channel and along the northwesterly trending spit that extends from South Island (Liu et al., this volume).

Summary

New Inlet and its backbarrier system are dynamic regions responding to a variety of processes brought about by the breaching of Nauset Spit. The most important of these have been greater tidal flow, caused by an increase in tidal range, and an influx of sand coming primarily from sediment moved across the swash platform. These conditions have resulted in a strengthening of the segregation of flood and ebb tidal currents and sand movement within Chatham Harbor. The flood ramp and flood-tidal delta are dominated by flood currents and northward sand transport. The doubling in size of the flood delta and the northward migration of the ebb shield (average = 150 m) from January 1987 to August 1990 are evidence of these processes. The adjacent channels are dominated by ebb flow and seaward sand transport. The increased discharge in East Channel has resulted in some deepening including the erosion and removal of the large, ebb-oriented, spillover lobe that extended from the southern end of the flood delta to two thirds of the way across East Channel prior to the breaching.

Backbarrier channels and shoals are dominated by medium-coarse sand with isolated pockets of coarse sand and gravel that coincide with deep channel regions having strong currents. The sands are mostly well-sorted to moderately well-sorted. Excepting a portion of the West Channel where currents are strong, the Chatham Harbor region is covered with large bedforms ranging in size from megaripples to sandwaves. Sandwaves are the most prevalent bedform in the channels having wavelengths of 16 to 43 m (average wavelength = 24 m) with heights of 0.5 to 3 m (average height = 1.1 m). Many of the sandwaves are complex forms with megaripples on their backs. The flood delta and other shoal areas are covered with sandwaves with a wider range in wavelength (8 to 69 m) and smaller heights (< 1 m).

The mouth of the inlet is separated into two opposing sediment transport zones: 1. a shallow swash platform, consisting of bars and channels, that is dominated by flood-tidal and wave-generated currents which produce landward-directed sand transport, and 2. a strongly ebb-tidally dominated inlet throat region where net seaward sediment transport predominates.

Acknowledgments

This study was supported by a contract (#DAC3990M4034) from the U.S. Army Corps of Engineers, Waterways Experiment Station, Vicksburg, MS. The authors wish to thank the Town of Chatham for dock space and the general assistance of the Harbormaster, Stu Smith. The inlet throat instrument package was deployed and retrieved by Mark Avakian of TG&B Consultants. Doug Levin, Steve Goodbred, and J.B. Smith are gratefully acknowledged for their assistance in the field. Paul Losquadro and David Byers performed the grain size analysis and Eliza McClennan prepared the figures.

References

Friedrichs, C.T., D.G. Aubrey, G.S. Giese and P.E. Speer, 1993. Hydrodynamical modeling of a multiple-inlet estuary/barrier system: Insight into tidal inlet formation and stability. In: Aubrey, D.G. and G.S. Giese, (eds.), *Formation and Evolution of Multiple Tidal Inlet Systems, Coastal and Estuarnie Studies,* AGU, (this volume).

Boothroyd, J.C. and D.K. Hubbard, 1975. Genesis of bedforms in mesotidal estuaries. In: Cronin, L.E. (ed.), *Estuarine Research,* New York, Academic Press, v. 2, p. 217-234.

Caston, V.N.D., 1972. Linear sand banks in the southern North Sea. *Sedimentology,* v. 18, p. 63-78.

Cornillon, P., 1979. Computer simulation of shoreline recession rates, Outer Cape Cod. In: Leatherman, S.P. (ed.), *Environmental Geological Guide to Cape Cod National Seashore,* Fieldtrip Guide Book for Eastern section of SEPM, p. 41-54.

FitzGerald, D.M. and T.M. Montello, 1990. A preliminary study of changes in bedform distribution and shoal morphology, sedimentation trends, and inlet hydraulics at New Inlet, Cape Cod, MA. Tech. Rpt. No. 14, Coastal Envir. Res. Group, Geology Dept., Boston University, Boston, MA, 90 pp.

Giese, G.S., 1988. Cyclic behavior of the tidal inlet at Nausct Beach, Chatham, MA. In: Aubrey, D.G. and L. Weishar, (eds.), *Hydrodynamics and Sediment Dynamics of Tidal Inlets,* Springer-Verlag, New York, p. 269-283.

Giese, G.S., 1990. The story behind the New Inlet at Chatham. *Nor'easter,* Magazine of Northeast Sea Grant Programs, v. 2, p. 28-33.

Hayes, M.O., 1975. Morphology of sand accumulation in estuaries: an introduction to the symposium In: Cronin, L.E. (ed.), *Estuarine Research,* New York, Academic Press, v. 2. p. 1-22.

Hine, A.C., 1972. Sand deposition in the Chatham Harbor Estuary and on the neighboring beaches, Cape Cod, MA. Unpub. MS Thesis, University of Massachusetts, Amherst, MA, 187 pp.

Hine, A.C., 1975. Bedform distribution and migration patterns on tidal deltas in the Chatham Harbor Estuary, Cape Cod, MA. In: Cronin, L.E. (ed.), *Estuarine Research,* New York, Academic Press, p. 235-252.

Liu, T.J., D.K. Stauble, G.S. Giese and D.G. Aubrey, 1993. Morphodynamic evolution of a newly formed tidal inlet. In: Aubrey, D.G. and G.S. Giese, (eds.), *Formation and Evolution of Multiple Tidal Inlet Systems, Coastal and Estuarnie Studies,* AGU, (this volume).

McClennen, C.E., 1979. Nauset Spit: a model of cyclical breaching and spit regeneration during a coastal retreat. In: Leatherman, S.P. (ed.), *Environmental Geological Guide to Cape Cod National Seashore*, Fieldtrip Guide Book for Eastern section of SEPM, p. 41-54.

Weidman, C.R. and J.R. Ebert, 1993. Cyclic spit morphology in a developing inlet system. In: Aubrey, D.G. and G.S. Giese, (eds.), *Formation and Evolution of Multiple Tidal Inlet Systems, Coastal and Estuarnie Studies*, AGU, (this volume).

6

Cyclic Spit Morphology in a Developing Inlet System

Christopher R. Weidman and James R. Ebert

Abstract

A spit attached to the north end of South Beach, a barrier island in Chatham, Massachusetts, exhibits a cyclic pattern of accretion, breaching, and shoal generation. The flood-oriented spit is within a large (2 km wide) developing tidal inlet system that was formed when Nauset Beach was breached in January 1987. Five cycles of spit growth and breaching have been observed in two years. Each cycle is characterized by: 1) elongation of spit for several months; 2) gradual narrowing and eventual breaching of the spit's mid-section, which creates a small tidal channel and terminates the growth cycle; and 3) evolution of the detached distal portion of the spit from a supratidal island to subtidal shoals. The growth and breaching of this spit is are significant processes causing landward sediment transport within the inlet system. A conceptual model is offered which explains the cyclic behavior of this system as the result of interrelated morphological, tidal, and climatic controls.

Introduction

The cyclic growth and breaching of barrier beaches and spits is a subject of great interest in coastal geology (DeBoers, 1964; Goldsmith, 1972; Ogden,

Formation and Evolution of Multiple Tidal Inlets
Coastal and Estuarine Studies, Volume 44, Pages 186-212
Copyright 1993 by the American Geophysical Union

1976; McClennen, 1979; Aubrey and Gaines, 1982; Nicholls, 1984; Giese, 1988). These periodic systems, once identified, can act as natural laboratories to improve our understanding of inlet-genesis and barrier beach/spit evolution. However, the investigational challenges imposed by the large spatial scales (1-10's km) and the long periodicities (10-100's yrs) of most previously described systems make it difficult to verify the morphological models derived from them. The modest physical scale of the spit system along the northern end of South Beach in Chatham, Massachusetts, and its frequent recurrence of spit elongation, breaching, and tidal inlet formation, provide an opportunity to overcome these earlier limitations.

The study area is located on the northern portion of South Beach, a barrier island approximately 1 km east of the mainland coast of Chatham, Massachusetts (Figs. 1 & 2). The island is 5 km long, 200-800 m wide, and has a north-south orientation. South Beach became an island when Nauset Beach was breached during a severe northeast storm that coincided with perigean tides on January 2, 1987. The breach's development shortened the path of tidal flow into and out of the Chatham Harbor and Pleasant Bay system (~25 km^2), and as a result the mean tidal range within Chatham Harbor increased from a pre-breach 1.8 m to a post-breach 2.1 m (U.S. Army Corps of Engineers, 1989). Prior to the 1987 breach, Nauset Beach extended 17 km south from its mainland attachment at Orleans and had been the focus of a number of morphological investigations (U.S. Army Corps of Engineers, 1957, 1968; Goldsmith, 1972; Hine, 1975, 1979; Giese, 1978; McClennen, 1979; and Hayes, 1981). Several studies (Goldsmith, 1972; Giese, 1978, 1988; McClennen, 1979) have outlined a cyclic history of spit accretion, inlet-migration, breaching, and deterioration of Nauset Beach opposite the Chatham mainland with a period of 100-150 years. As part of this larger cycle, South Beach, the deposits of which represent the last 50 yrs of Nauset Beach's growth (Hayes, 1981), is expected to erode during the next several decades, with its sediments moving southward and westward (Giese, 1988).

Since the inception of the new inlet through Nauset Beach, flood-oriented spits have formed on both sides of the inlet (Fig. 3). The term "inlet-spit" was coined (Weidman and Ebert, 1988; Ebert and Weidman, 1989) to distinguish these recent, unvegetated, and rapidly changing depositional features from the older, vegetated, and relatively stable barrier systems to which they are attached. The morphology of the inlet-spits in Chatham must be considered

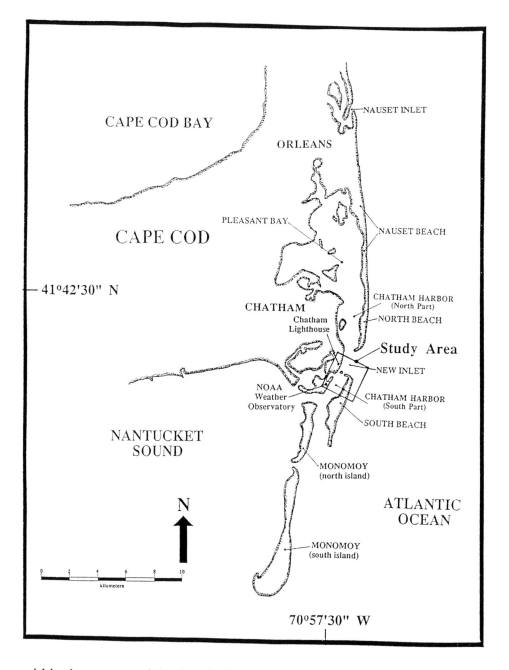

within the context of the developing inlet system which widened from less than 100 m to over 2000 m from January 1987 to January 1990. This widening is part of the continuing adjustment of the system to the greater hydraulic efficiency and greater tidal prism afforded by the new inlet. The enlargement of the inlet is produced by: 1) tidal currents which scour and deepen the inlet; and 2) waves which erode the inlet's margins. These processes provide an abundant sediment supply which allows for the construction of inlet-spits. At the same time, tidal currents and wave action provide a basis for the inlet-spits' unstable existence by eroding their inlet-facing beaches and causing the inlet-spits to narrow even as they grow in length.

The inlet-spits on both sides of the inlet have displayed a repeating pattern of accretion and breaching. The inlet-spit attached to North Beach breached twice in two years, whereas the inlet-spit attached to South Beach breached five times during the same period. This difference may reflect the asymmetry in the sedimentological and hydrodynamic processes affecting the New Inlet system. The orientation of South Beach leaves its northern end vulnerable to

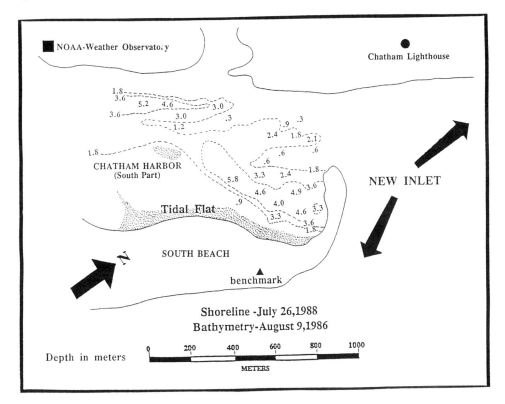

Figure 2. Map of the study area on the north end of South Beach. The spit and barrier shoreline are from a post-breach (1987) survey (on July 26, 1988), while the harbor bathymetry is pre-breach (from NOAA-National Ocean Service Chart #13248, corrected to August 9, 1986). Note the deepest parts of the harbor adjacent to the spit are opposite the spit's mid-section.

the dominant storm wave approach (northeast), whereas the orientation of North Beach's southern end is relatively sheltered during these same storms. Further, North Beach is the present distal part of Nauset Beach and its sediment is derived from longshore transport along the 15 km of shoreline south of Nauset Inlet, whereas South Beach is down-drift of the new inlet and its sediment supply is derived only from sediment eroded from the northern end of South Beach and from sediment able to bypass the inlet.

Figure 3. Aerial survey photograph (September 1, 1988) of the New Inlet area in Chatham showing the inlet-spits attached to the older barriers on both sides of the inlet. Note the recently-formed breach channel in the mid-section of the inlet-spit attached to South Beach. The square-ish shoal to the upper left of the detached distal end of the South Beach inlet-spit is also a former detached distal end from an earlier (January 1988) breach.

This paper documents five cycles of growth and breaching of the inlet-spit attached to the northern end of South Beach, and confirms that the inlet-spit is an important avenue of landward sediment transport within the inlet system. The observations are used to develop a conceptual model of inlet-spit evolution which explains the cyclic morphology of this feature as a product of morphological, tidal, and meteorological controls. Components of the model, supported by previous studies of barrier and spit evolution, can also provide some insight to the general problem of barrier and spit breaching.

Methods

Fieldwork began on November 14-15, 1987, and a benchmark (for horizontal control) was established on the northern end of South Beach from which to repeat surveys (Fig. 2). Barrier scarpline retreat on the north end of South Beach was measured from fixed markers placed every 30 m in a series of lines parallel to the axis of South Beach. Since the inlet-spit was subject to overwash and breaching, it was resurveyed during each field visit by running a baseline of temporary markers (spaced every 30 m) along the axis of the spit and tied into the benchmark on South Beach. Cross-sectional distances from the baseline markers on the spit's axis to the toe of the spit's foreshore and backshore were measured at low tide. The detached distal ends of the inlet-spit were surveyed in this same manner, but the accuracy of their shoreline positions is less certain than the shoreline positions of the spit because of their isolation from the benchmark on South Beach. These measurements were used to construct a map of the inlet-spit's low tide shoreline and the adjacent barrier's scarpline for each field visit. Survey schedules were dictated largely by convenience, though efforts were made to resurvey as soon as possible after a breach was reported by local observers. The durations of intervals between successive surveys varied from 1-9 weeks with an average interval duration of 30 days.

Hourly weather summaries were acquired from the NOAA Weather Observatory-Chatham Station, located 1.5 km from the study area (Fig. 2). The hourly data were sub-sampled to obtain the wind velocities for only the 3 hrs before and the 3 hrs after the time of the forecast high tides (Tide Tables, 1987, 1988, 1989). These 6-hr blocks of wind data were then reduced to a single resultant wind velocity vector for each high tide (referred to as the "high-tide-wind"). This method was chosen in order to emphasize wind conditions at the time of high tide when storm-induced changes to the inlet-spit would be maximized, and to de-emphasize wind conditions at the time of low tide when these changes would be minimized. The "high-tide-winds" were compiled for each of the intervals between successive surveys to produce two indices which were relevant to the morphology of the inlet-spit: 1) storm frequency — the percentage of an interval's "high-tide-winds" with onshore direction (0°- 200°) and speeds > 10.0 kts (18.5 km/hr); 2) storm intensity — the mean wind speed of these stormy "high-tide-winds" (stormy being defined in the same manner as the storm frequency index). Indices derived in this way have been successfully used previously to model scarp erosion on South Beach (Weidman, 1988; Weidman and Ebert, 1988, 1989).

Results

Two years of field mapping document five cycles of inlet-spit growth, breaching, and shoal generation. Consecutive shoreline positions of the inlet-spit and the adjacent barrier scarpline are presented along with morphological and meteorological characterizations for each interval between surveys (Figs. 4-8). These interval characterizations include: survey dates; duration; barrier scarpline erosion rate (linear retreat rate at the spit base/barrier boundary); spit growth rate (length and area); and storm indices (frequency and intensity). Intervals are grouped according to cycle, where the term "cycle" is defined as the period of inlet-spit growth between successive breachings. Breachings generally resulted in the formation of a subtidal breach channel, causing a significant shortening of the spit and detachment of its distal end. Composites of each of the five inlet-spit growth cycles are presented along with each cycle's morphological and meteorological summary (Fig. 9). The summaries do not include intervals in which breaches occurred, since during these intervals' morphological changes could not be ascribed to a particular cycle. For the entire two-year (707 days) period of study, 330 meters were eroded from the northern end of South Beach for a mean barrier erosion rate of 0.5 m/day, and inlet-spit growth rates averaged 1.9 m/day for length and 160 m²/day for area. The sediment volumes transported into Chatham Harbor from the growth of the inlet-spit on South Beach can be roughly (and conservatively) calculated if the thickness of the inlet-spit deposits are assumed to be about the same as the tidal range (~2 m) and subtidal volume and beach slope effects are neglected. The results of this calculation over the duration of this study is a total transported sediment volume of ~210,000 m³ or an average of ~300 m³/day. Storm frequency and storm intensity indicies for the entire study were 16.8% and 26.6 km/hr respectively.

Cycle I

Documentation of Cycle I (Fig. 4) is incomplete since the investigation began with the first cycle already in progress. Oblique air photos (Kelsey-Kennard Airviews, 1987) indicate a prior breach occurring sometime in mid-June 1987, and the spit is assumed to have grown uninterrupted throughout the summer and fall of 1987. The north end of South Beach and the base of the inlet spit eroded much faster than the study's mean rate during interval 1 and about the same as the mean rate during interval 2. Inlet-spit elongation was much faster during interval 1 than during interval 2, though the areal growth rate was nearly three times as great during interval 2 as during interval 1. Distal accretion was oriented to the west for both of the intervals — giving the spit a pronounced recurve. During interval 1, the mid-section narrowed to 50 m and a washover fan prograded on the bay side of the proximal spit during interval 2. Intervals 1 and 2 were both characterized by moderate (average) storm frequencies, whereas the storm intensity index was highest (and well above the mean) during interval 2.

The inlet-spit was reported to have breached on January 19-20, 1988, ending a 7-month long cycle of inlet-spit growth. The interval in which the breach occurred was characterized by low storm frequency and storm intensity indices. The breach coincided with peak spring tides and moderate intensity onshore (southeast) winds. The breach channel was just over 100 m wide and its center was located about 250 m from the base of the spit when it was surveyed 11 days later on January 31, 1988. The remnant spit was 200 m long, 50 m wide, and its axis had rotated 15° counterclockwise from its previous orientation. The detached distal portion of the spit was a supratidal island, 170 m long and 100 m wide.

INLET-SPIT CYCLE I

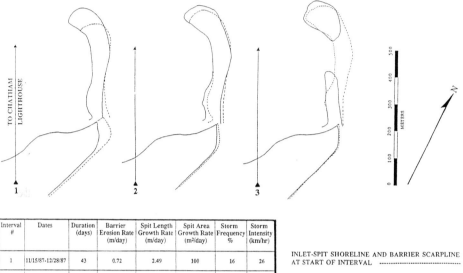

Interval #	Dates	Duration (days)	Barrier Erosion Rate (m/day)	Spit Length Growth Rate (m/day)	Spit Area Growth Rate (m²/day)	Storm Frequency %	Storm Intensity (km/hr)
1	11/15/87-12/28/87	43	0.72	2.49	100	16	26
2	12/28/87-1/12/88	15	0.43	1.80	287	14	32
3	1/12/88-1/31/88	19	0.39	BREACH	BREACH	11	22

INLET-SPIT SHORELINE AND BARRIER SCARPLINE
AT START OF INTERVAL   --------------------------------

INLET-SPIT SHORELINE AND BARRIER SCARPLINE
AT END OF INTERVAL   _____

Figure 4. Inlet-spit Cycle I: showing successive shoreline and barrier scarpline positions for intervals #1-3 along with morphological and meteorological characterizations for each interval. This cycle's documentation is incomplete since the study began several months into the spit's growth cycle. The spit was reported breached on January 19-201, 1988.

Cycle II

Intervals 4-8 (Fig. 5) comprise the study's first completely documented cycle of inlet-spit growth. Less than 3 months in duration, it was the briefest of the five cycles and was fully within a winter climate. Barrier erosion rates were generally above the mean throughout the cycle. During the stormier intervals (4, 6, 8), the spit shifted laterally to the west via overwash. Spit growth (length and area) was slow during these same intervals. Elongation of the spit was greatest during interval 5 and 7, which had the cycle's milder storm intensities.

The detached distal section (from the January 1988 breach) eroded on its eastward side and these sediments were redeposited on its bay side causing the island to migrate westward. The apparent increase in the island's area as this migration occurred was the result of an overall lowering and spreading of the island's sediments from wave action and overwash. In this way, the island's supratidal area was gradually reduced, and it became an intertidal shoal during the same week that the spit was breached.

On April 6th, the spit was a modest 310 m long and a narrow 50-60 m wide. Overwash of the spit's mid-section was observed during this field visit, which had a "high tide wind" of 13.2 kts (24.5 km/hr) and 13°(~NNE). For about an hour at high tide, swash bores crossed a ~50 m wide swath of the spit's mid-section from the inlet side to the bay side. The inlet-spit was breached sometime during the following interval 8, which was only one week long and characterized by a high storm frequency index and a moderate storm intensity index. Though the interval coincided with neap tides, the constant storm conditions were apparently sufficient to cause a breach in the mid-section about 230 m from the base of the spit. Only a small distal section was detached

INLET-SPIT CYCLE II

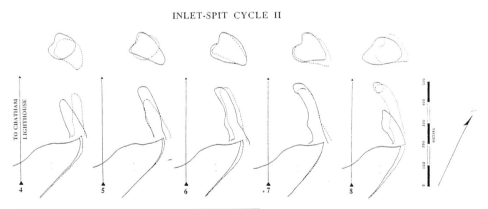

Interval #	Dates	Duration (days)	Barrier Erosion Rate (m/day)	Spit Length Growth Rate (m/day)	Spit Area Growth Rate (m²/day)	Storm Frequency %	Storm Intensity (km/hr)
4	1/31/88-2/25/88	25	0.59	-0.28	108	14	30
5	2/25/88-3/12/88	16	0.61	4.38	213	10	22
6	3/12/88-3/28/88	16	0.76	0.81	81	16	30
7	3/28/88-4/6/88	9	0.00	4.33	111	17	24
8	4/6/88-4/13/88	7	3.48	BREACH	BREACH	83	26

INLET-SPIT SHORELINE AND BARRIER SCARPLINE
AT START OF INTERVAL ------------------------------

INLET-SPIT SHORELINE AND BARRIER SCARPLINE
AT END OF INTERVAL ─────────────────

Figure 5. Inlet-spit Cycle II: showing successive shoreline and barrier scarpline positions for intervals 4-8 along with morphological and meteorological characterizations for each interval. This is the briefest of the five cycles with a short spit length at the time of breaching. Note the westward migration of the detached distal section.

because of the abbreviated length of the spit, and when the spit was surveyed on April 13th, the detached distal section was an intertidal shoal. The remant spit was 180 m long, 70 m wide, and its axis had rotated 10° counterclockwise from its previous orientation.

Cycle III

Intervals 9-12 (Fig. 6) comprise a 4-1/2 month-long cycle. Storm frequencies and intensities decreased for the first three intervals, reflecting a seasonal passage from early spring to summer. The overall barrier erosion rate for the cycle was near the study's mean, whereas growth rates for the spit were the highest for all cycles. Erosion of the adjacent barrier was well below the mean during the mild intervals 10 and 11, and above the mean during the stormy intervals 9 and 12. During interval 9, the spit's bay side shoreline prograded and elongation of the spit was most rapid.

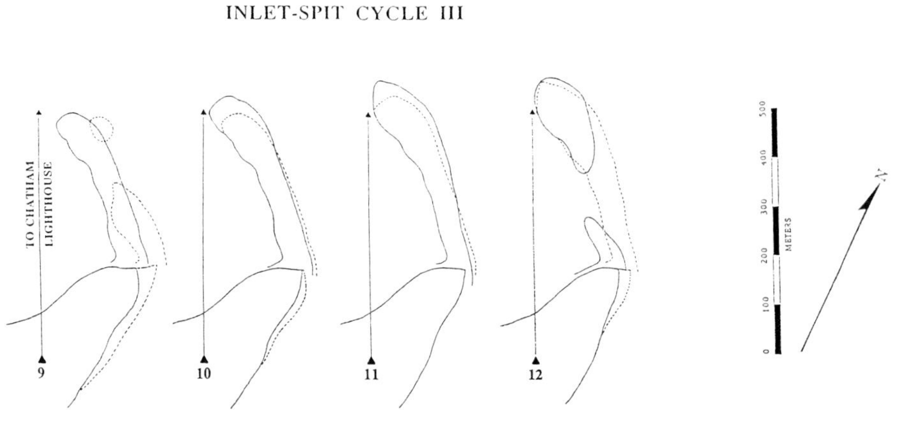

INLET-SPIT CYCLE III

Interval #	Dates	Duration (days)	Barrier Erosion Rate (m/day)	Spit Length Growth Rate (m/day)	Spit Area Growth Rate (m²/day)	Storm Frequency %	Storm Intensity (km/hr)
9	4/13/88-6/7/88	55	0.83	3.04	198	21	24
10	6/7/88-7/5/88	28	0.16	1.57	96	9	24
11	7/5/88-7/26/88	21	0.00	1.95	214	7	20
12	7/26/88-8/29/88	34	0.75	BREACH	BREACH	14	24

INLET-SPIT SHORELINE AND BARRIER SCARPLINE AT START OF INTERVAL ························

INLET-SPIT SHORELINE AND BARRIER SCARPLINE AT END OF INTERVAL ────────

Figure 6. Inlet-spit Cycle III: showing successive shoreline and barrier scarpline positions for intervals 9-12 along with morphological and meteorological characterizations for each interval. Note the oceanward shift of the transition point from erosion to accretion on the inlet-spit between interval 10 and interval 11 as conditions became milder. The spit was breached on August 27-28, 1988.

During intervals 10 and 11, the distal end widened by accretion on the inlet-side beachface. This lateral accretion combined with the proximal beachface erosion caused a clockwise rotation of the spit axis. No overwash occurred during these mild intervals and the proximal- and mid-section narrowed to 50 m.

The spit was about 420 m long when the third breach was reported on August 27, 1988, during peak spring tides coincident with moderate onshore storm winds. Two days later on August 29th, the breach channel was measured to be 90 m wide and was centered about 170 m from the spit's base. The remnant spit was 110 m long, 40 m wide, and its axis had rotated 17°counterclockwise from its previous orientation.

Cycle IV

Intervals 13-17 (Fig. 7) comprise a 6-to-7 month long cycle. The ambiguity in cycle duration is due to the uncertainty of the timing of the breach during interval 17. This cycle was the "stormiest" of the five cycles with storm frequencies and intensities generally above the mean, reflecting a seasonal passage from early fall to late winter. Barrier erosion was greatest during interval 14 and 17, the stormiest intervals. Elongation of the spit was also slowest during stormy interval 14, but most rapid during interval 16 which had an average storm frequency and a high storm intensity. Bay side progradation of the spit via overwash occurred during intervals 14 and 15. The spit's mid-section narrowed to less than 60 m during interval 16, the last interval before breaching.

The date of the fourth breach is uncertain, but it occurred sometime between February 20-March 15, 1989. The meteorological record suggests two triggering events, a 2-day northeaster on February 23-25th, and a 4-day northeaster on March 6-10th. Tide Table forecasts indicate the latter storm as the triggering event since it coincided with peak spring tides. The breach channel (measured at least 6 weeks later) was centered about 320 m from the base of the spit and was 230 m wide. The size of the detached distal section near the time of the breach is unknown, but a small supratidal island was surveyed on April 23, 1989. The remant spit (at least 6 weeks into its growth

INLET-SPIT CYCLE IV

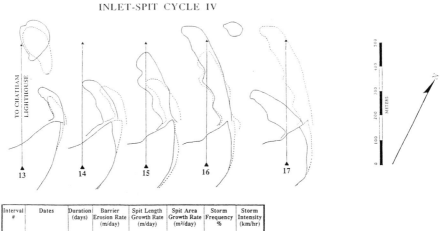

Interval #	Dates	Duration (days)	Barrier Erosion Rate (m/day)	Spit Length Growth Rate (m/day)	Spit Area Growth Rate (m²/day)	Storm Frequency %	Storm Intensity (km/hr)
13	8/29/88-9/26/88	28	0.42	1.57	125	19	24
14	9/26/88-10/31/88	35	1.37	1.09	129	27	30
15	10/31/88-12/28/88	58	0.11	1.90	195	18	28
16	12/28/88-2/18/89	52	0.53	2.79	154	17	32
17	2/18/89-4/23/89	64	0.76	BREACH	BREACH	26	32

INLET-SPIT SHORELINE AND BARRIER SCARPLINE
AT START OF INTERVAL ··················

INLET-SPIT SHORELINE AND BARRIER SCARPLINE
AT END OF INTERVAL ————————

Figure 7. Inlet-spit Cycle IV: showing successive shoreline and barrier scarpline positions for intervals 13-17 along with morphological and meteorological characterizations for each interval. This is the stormiest of the five cycles. Interval 14, one of the stormiest intervals, has the cycle's highest barrier erosion rate and the slowest growth rate of the inlet-spit. The spit was breached sometime between February 20 - March 15, 1989.

cycle) was 240 m long, 70 m wide, and its axis had rotated 32° counterclock-wise from its previous orientation.

Cycle V

The fifth cycle (Fig. 8) began during interval 17 and extended at least seven months until late October, 1989, when a breach occurred in the distal section of the spit. This cycle was the mildest of the five cycles and characterized by low storm frequencies and intensities during the spring and summer intervals and by moderate-to-high storm frequencies and intensities during the later fall periods. The cycle's barrier scarpline erosion rate was practically negligible. The elongation rate was the least of the five cycles, though the areal growth rate was near the mean. During most of the cycle's intervals, the entire spit-beachface accreted and the mid-section eventually widened to more than

INLET-SPIT CYCLE V

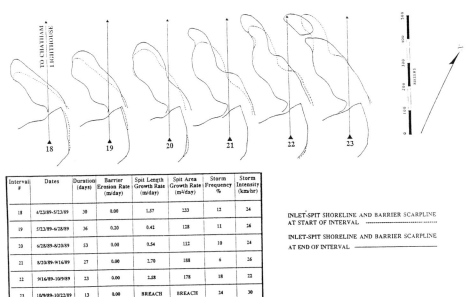

Interval #	Dates	Duration (days)	Barrier Erosion Rate (m/day)	Spit Length Growth Rate (m/day)	Spit Area Growth Rate (m²/day)	Storm Frequency %	Storm Intensity (km/hr)
18	4/23/89-5/23/89	30	0.00	1.57	233	12	24
19	5/23/89-6/28/89	36	0.20	0.42	128	11	26
20	6/28/89-8/20/89	53	0.00	0.54	112	10	24
21	8/20/89-9/16/89	27	0.00	2.70	188	6	26
22	9/16/89-10/9/89	23	0.00	2.58	178	18	22
23	10/9/89-10/22/89	13	0.00	BREACH	BREACH	24	30

INLET-SPIT SHORELINE AND BARRIER SCARPLINE
AT START OF INTERVAL   ·····································

INLET-SPIT SHORELINE AND BARRIER SCARPLINE
AT END OF INTERVAL   ————————————

Figure 8. Inlet-spit Cycle V: showing successive shoreline and barrier scarpline positions for intervals 18-23 along with morphological and meteorological characterizations for each interval. This cycle has the lowest storm frequencies. Barrier erosion is negligible and the entire spit widens. The spit's distal section was breached on October 18-20, 1989.

100 m. The spit axis rotated clockwise as distal beachface accretion was greater than proximal accretion, however the spit axis still retained a pronounced westward orientation throughout the cycle. During intervals 20-23, a large lobe of sediment migrated downspit along the spit-beachface. Its advance was preceded by an area of erosion, causing an embayment which propagated downspit as well.

The inlet-spit was breached during a 3-day northeast storm (October 18-20, 1989) coincident with spring tides. At this time, the embayment was opposite a narrow portion of the distal section about 350 m from the base of the spit, and was likely an important factor in the spit's breaching at this location. The detached section was a small intertidal shoal when measured a few days later, and the breach channel was only 20 m wide. The remnant spit was 340 m long.

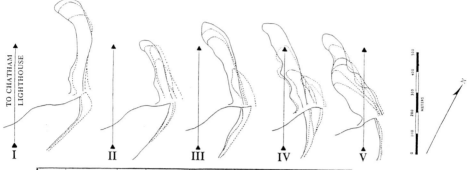

Figure 9. Composites of inlet-spit growth cycles along with morphological and meteorological summaries for each cycle. The summaries do not include data from intervals in which breaches occurred (intervals 3, 8, 12, 17 and 23), because morphological changes during these intervals could not be ascribed to a particular cycle. The width of the barrier increases as the north end of South Beach erodes. The inlet-spit rotates clockwise as it grows owing to erosion on its prosximal- and mid-sections while it accretes on its distal end.

This last breach did not fully terminate a cycle as defined above, but was a temporary interruption in a continuing cycle. The breach did not significantly shorten the inlet-spit and the detached distal portion was only a small intertidal shoal. The small breach channel was submerged for only part of the tide and was closed soon after the last survey by a resumption of spit growth.

Discussion

The morphology of inlet-spits can be discussed in the context of four processes:

Barrier and Proximal-Spit Erosion

The erosion of the South Beach barrier has been the most critical of the processes affecting the cycle of spit growth and breaching. The erosion of the barrier from wave action and tidal currents supplies the sediments that form the spits, though some sediment may also be derived from other sources to the north of the inlet, and from scouring processes within the inlet. At the same time, the erosion narrows the proximal- and mid-section of the spit which eventually allows for overwash and breaching. The barrier erosion rate is generally correlated with an interval's storminess and has been modeled previously as a function primarily of storm frequency (Weidman, 1988; Weidman and Ebert, 1989). However, there has been an overall decrease in the rate of barrier retreat on South Beach that cannot be fully explained by variations in storm frequency and intensity. Another control on barrier retreat may be the width of the South Beach's north end. The breach through Nauset Beach in 1987 cut through one of the narrowest sections of the barrier, and the inlet has been eroding the edges of successively wider regions of the barrier ever since, possibly resulting in an increased resistance to longitudinal retreat. Reduced barrier erosion may also indicate that the adjacent inlet has reached a state of equilibrium after 3 years of rapid development.

Distal-Spit Accretion

Distal accretion occurred both as elongation and as widening. Spit growth is often defined as elongation or axial growth, and this occurred in almost all non-breaching intervals, both stormy and mild. No obvious correlation exists between an interval's rate of axial growth and its barrier erosion rate, storm frequency, or storm intensity. However, when storms were severe enough to cause pervasive overwashing (intervals 4, 6, and 14), the spits simply translated bayward and elongation was slowed or did not occur. Climate was also qualitatively reflected in the orientation of the axial accretion, which was oriented to the west during stormier intervals and to the north during milder intervals. Axial growth of the spit narrows the bay entrance between the distal spit and the mainland. There is no clear evidence that axial growth rates were more constrained as the spits grew longer despite the potential for increased tidal velocities through this entrance. Adding to the complexity, the erosion and migration of the detached distal ends caused a progressive shoaling of the bay entrance during the course of this study. Though a shallower bathymetry could enhance spit elongation, a deep tidal channel was continually scoured directly ahead of the distal spit (Fig. 3) and much of the pre-existing bathymetry was reworked before distal deposition took place. Widening of the distal spit or neap-berm building in the manner described by Hine (1979) was common during mild intervals and absent during the stormy intervals.

Overwash Progradation

Progradation of the spit's bayside shoreline was accomplished by overwash processes during storms. The proximal and mid-section of the inlet-spits were the regions narrow enough to allow overwash to traverse the spit. The bathymetry adjacent to the bay side of the spit determined the configuration of the overwash-derived sediment volume. The shallow tidal-flat opposite the proximal section (within 100 m of the base) allowed overwashed sediments to prograde more rapidly than did the deeper regions opposite the mid-section (150-250 m beyond the base).

Breaching

Breaching occurred when all or most of the following conditions existed: 1) storms with onshore winds (probably all cycles, but the circumstances of the fourth breach are uncertain); 2) spring tides (perhaps four out of five cycles, but again the circumstances of the fourth breach are uncertain); 3) a spit length in excess of 400 m (four out of five cycles); 4) a minimum spit width of 60 m (all cycles); this location was usually the mid-section 150-250 m beyond the base of the spit.

Model

Based on these observations, a conceptual model has been constructed of the

CONCEPTUAL MODEL
OF
INLET-SPIT GROWTH CYCLE

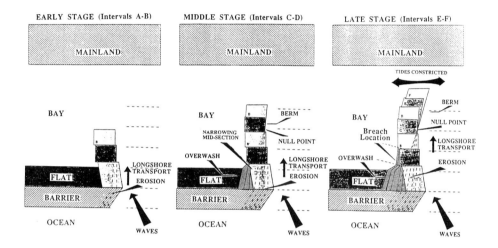

Figure 10. A conceptual model of inlet-spit evolution is illustrated in three stages of development. Breaching of the spit in the late stage terminates one cycle and begins the early stage of a new cycle. Not shown above is that following breaching (during the early stage of the new cycle) the detached distal section erodes and its sediments move bayward and towards the mainland. The remnant spit's axis rotates counterclockwise from its previous orientation and spit growth resumes.

morphological evolution of the South Beach inlet-spit. This model is illustrated in three stages of inlet-spit evolution for one cycle (Fig. 10).

Early Stage (Intervals A-B)

Waves and tidal currents erode the narrow end of the barrier adjacent to the inlet. The short spit elongates as longshore transport supplies sediments derived from the eroding barrier and spit-beachface. The proximal section narrows as the spit-beachface erodes.

Middle Stage (Intervals C-D)

Erosion of the barrier and the spit's proximal beachface decreases downspit, reflecting a bayward decrease in wave energy. Overwash crosses the spit as the proximal section is critically narrowed forming a prograding washover fan on the adjacent back-barrier tidal flat. In this way, the proximal section is able to maintain its sub-aerial width despite beachface losses. Progradation of the bay side shoreline is retarded in the deeper water beyond the tidal-flat and so the mid-section narrows. Elongation extends the distal portion of the spit beyond the null point where net erosion yields to net accretion.

Late Stage (Intervals E-F)

The spit's mid-section becomes critically narrow as the spit's beachface continues to erode. Distal beachface accretion occurs beyond the null point in the form of successive berms, and the distal spit widens as a result. As the spit elongates, the separation between it and the mainland narrows which causes an increasing constriction of the tidal flow into and out of the bay. This tidal constriction causes a decreasing tidal amplitude in the bay and an increasing lag in the bay's tidal phase relative to the ocean tide. This results in an increasing hydraulic head across the spit. The hydraulic head gradient (hydraulic head/ width of spit) will be greatest at the location where the spit is narrowest — the mid-section. The morphology of the inlet-spit is now conducive to breaching under suitable tidal and meteorological conditions. Following breaching, the model reverts back to the early stage and the cycle repeats. The remnant spit is reoriented counterclockwise owing to the dominant wave conditions on the inlet side of the spit and the new tidal entrance's closer proximity to the ocean. The detached distal section erodes as its sediment supply is interrupted by the formation of the breach channel, and this island/shoal migrates bayward in response to the dominant wave conditions on its oceanward side.

The main aspects of the conceptual model are supported by comparing it (Fig. 10) with the composite time series of inlet-spit growth cycles (Fig. 9). During most of the growth cycles, the spit's axis appears to rotate clockwise around a node located in its mid-section, about 200 m beyond its base. This apparent rotation is caused by the increasing erosion oceanward of the node and the increasing beachface accretion bayward of the node. The proximal section maintains its width, whereas the mid-section generally narrows, with the breaches occurring in the mid-section of the inlet-spit.

Climatic Imprint on Inlet-Spit Morphology

The variation in spit morphology between one cycle and the next can largely be explained by each cycle's climatic history. The small scale, rapid growth, and short duration of the inlet-spit cycles on South Beach allow synoptic and seasonal changes in storm climate to be strongly correlated with the morphology of the spit. Surprisingly, there is no evidence from the field data that sediment supply or inlet-spit growth (elongation or areal growth) is related to energy conditions. This may be because northeast storms, while effective agents of erosion of the north end of South Beach and the proximal sections of the spit, are also effective at depositing this eroded material elsewhere besides the inlet-spit. This material may be deposited farther into the bay, on the other shoals within the lagoon and inlet, or may be moved south along the outside of South Beach. Fairweather conditions at Chatham, have a dominant southerly component (Weidman, 1988) and may move sediment northward along the outside of South Beach, into the inlet, and along the inlet-spit. Also, the reduced energy conditions may tend to inhibit the transport of sediment off the inlet-spit. In this way, compensating processes appear to provide a more or less steady supply of sediment to the inlet-spit under contrasting energy conditions. However, different wave energy conditions (stormy vs. mild) do cause characteristic erosion and accretion, producing spits with different orientations and shapes (Fig. 11). This can be demonstrated by comparing the morphologies and climates of each of the five cycles .

Cycle I

The strong recurve of the inlet-spit's distal end is explained by increasing wave energy conditions as the spit grew. This is consistent with a cycle that began in early summer and ended in mid-winter. The earlier mild conditions

CLIMATIC IMPRINT
ON
SPIT MORPHOLOGY

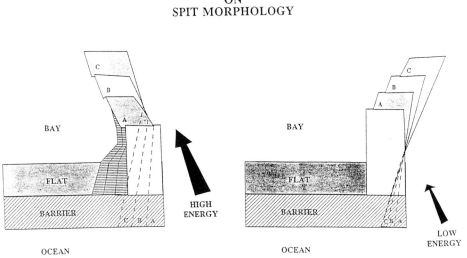

Figure 11. The influence of climate on inlet-spit morphology is most apparent in the orientation of its distal accretion.

resulted in a rather straight, nearly northward-oriented proximal section and mid-section (indicating some lateral beachface accretion). The later stormier conditions caused the sediment to be deposited farther westward around the spit's distal end, causing the recurve.

Cycle II

The westward orientation of the inlet-spit, the narrowness of its distal portion, and the brief duration of its cycle are all evidence of stormy conditions, consistent with a cycle entirely within a winter climate. The greater energy conditions erode the spit-beachface and cause frequent overwashing, resulting in a narrow spit that is subject to early breaching.

Cycle III

The spit's wide distal end and narrow mid-section reflect distal beachface (lateral) accretion and proximal beachface erosion. This was a product of increasingly mild conditions characteristic of a passage from spring to summer. Except for a lack of overwash deposition after the first interval of the cycle, this cycle is most similar to the model "type".

Cycle IV

The spit's narrowness throughout its length is evidence of stormy conditions throughout its cycle, consistent with a fall to late winter climate. The early intervals of this cycle are reminscent of cycle II, another winter cycle, when beachface erosion and overwash caused the bayward migration of the entire spit. The spit attained a long length despite a narrow mid-section and persistent storminess for most of its cycle. The meteorological and tidal record explain this by a lack of coincident storms and peak tidal conditions, so that a breach was not triggered for many weeks despite the inlet-spit's morphological "ripeness" for breaching.

Cycle V

An anomolously wide spit without a narrow mid-section was produced during a cycle characterized by little erosion of the barrier and proximal spit-beachface, and steady distal spit-beachface accretion. The extreme westward orientation of the initial spit was a product of stormy (late winter) conditions which existed at the beginning of the cycle. This was followed by 5 months of mild conditions (spring and summer). This sequence of energy conditions allowed for a relatively large clockwise rotation of the spit axis and a wide mid-section. The narrow distal portion in the later intervals reflects increasing storminess.

Inlet-Spit as Scale Model

The short period cyclicity of the South Beach inlet-spits provides an opportunity to understand better the evolution of spit systems that are larger in
spatial and temporal scale. As a precaution, it should be recognized that the
inlet-spit on South Beach exists within the unique environment of a developing tidal inlet. However, some previous investigations of larger scale spit
accretion and breaching cycles reveal that these cycles conform to a number
of aspects of inlet-spit morphology and behavior.

DeBoer's study (1967) of the morphological history of Spurn Head in
Yorkshire, U.K., documents four cycles of spit growth and breaching from
almost a thousand years of historical record. He offers a model in which the
eroding mainland supplies sediment for spit elongation and simultaneously
erodes the proximal parts of the spit. This leads to a reorientation of the spit's
shoreline towards the dominant storm wave direction which in turn allows for
more frequent overwashing of the spit during storms. The mid-section
narrows through continued erosion and is eventually breached. This ends one
cycle and another begins.

The cyclic growth and breaching of Nauset Beach, the parent body of the
South Beach inlet-spits, is thought to be controlled by increased constriction
of the tides caused by the southward growth of the spit (Giese, 1978, 1988;
McClennen, 1979). According to Giese, elongation of the Nauset Beach spit
produces an increasing tidal phase lag and amplitude difference between the
Atlantic Ocean and Pleasant Bay. These tidal differences eventually reach a
critical stage, where a breach of the spit north of the extant inlet results in the
formation of a new permanent inlet, and the cycle repeats itself.

The morphological evolution of these systems is similar to that of the inlet-
spit on South Beach. All these spit systems evolve to where breaching
becomes inevitable. Their behavior supports a generalized concept of cyclic
spit morphology in which spit growth is characterized by elongation, reorientation, and thinning. Elongation constricts lagoonal tides and raises a
hydraulic head across the spit. Reorientation may increase the spit's exposure
to wave energy leading to more frequent overwash and beachface erosion.
The local storm climate and surrounding hydrodynamics determine the

critical width necessary for the spit to breach. A similar concept of critical barrier width as a prerequisite for washover fan progradation has been previously discussed by Leatherman (1979). The location of a breach is strongly influenced by the bay side bathymetry. Pierce (1970) argued that the barrier sections most susceptible to breaching are narrow and lack shallow tidal flats on their lagoon sides. Finally, storms whose effects may be enhanced by extreme tides provide the necessary conditions for triggering a breach. Wood (1976) has correlated significant coastal changes with coincident storms and extreme spring tides.

Conclusions

The South Beach inlet-spit displays a cyclic morphology of accretion and breaching. Five cycles have been documented in two years. This cycle is characterized by: 1) a 3-7 month period of spit elongation from a minimum length of 100 m to maximum length of 500 m; 2) a clockwise rotation of the spit axis reflecting a combination of distal spit-beachface accretion and proximal spit-beachface erosion; 3) gradual narrowing and the eventual breaching of the spit's mid-section 100-200 m from the spit's base, resulting in the formation of a small tidal channel between former spit-sections; and 4) evolution of the detached distal section from a supratidal island to subtidal shoals.

A conceptual model has been offered which explains the cyclic behavior of the inlet-spits as a consequence of interrelated of morphological, tidal, and meteorological factors. 1) The erosion of South Beach's northern end is forced by wave and tidal conditions related to the developing New Inlet system. 2) This erosion plays a dual role by supplying sediments for the construction of inlet-spits and simultaneously causing their destruction by narrowing their proximal- and mid-sections. 3) The bathymetry on the bay side of the spit constrains the width of the spit by influencing the rate of bay side progradation from overwash. The shallow regions adjacent to the proximal spit help to maintain the spit's width, while the deeper regions adjacent to the spit's mid-section allow it to narrow. 4) As the spit elongates, the distance between the spit and the mainland shortens, possibly constricting the tidal flow between the bay and the inlet. This would be reflected by increasing tidal phase and amplitude differences between the bay and inlet side. 5) The morphology of the spit is "ripe" for breaching when the spit is long (400-500 m), the mid-section is narrow (< 60 m), and consequently the hydraulic head gradient is greatest. 6) At this point, a sufficient combination of storm waves, storm surge, and spring tides will trigger a breach of the spit and initiate a small tidal channel, thus ending the cycle.

Differences between cycle morphologies are largely governed by their climatic histories. Cycle periods are measured in months, and so inlet-spits can complete their entire growth cycles within one or two seasons. The sequence of changes in storm frequency and intensity (and resulting wave energies) on a seasonal basis, causes characteristic patterns of erosion and accretion during spit growth. Mild weather favors beachface accretion and stormy weather favors beachface erosion and overwash. In this way, the inlet-spits retain a signature of their cycle's climatic history in their resultant shape and orientation.

Some aspects of inlet-spit morphology might be applied towards understanding the evolution of barrier and spit systems in general. The concept of a critical barrier width for breaching, as determined by the surrounding hydrodynamic conditions (both wind-wave- and tide-related), could be instrumental in assessing a system's stability. Lagoon bathymetry can play an important role in determining breach locations by affecting the rate of back-barrier migration. Seasonal or climatic changes may be reflected in the changing orientation of spits and barriers along their length and could be useful in reconstructing a climatic history of a region. Finally, it is important to emphasize that although storms trigger barrier breaching, the evolution of a system incorporates somewhat predictable hydrodynamic and morphological changes that can eventually produce a morphology that is "ripe" for breaching.

Acknowledgments

This study was partially funded by the Town of Chatham, the Commonwealth of Massachusetts through the Department of Environmental Management, the Coastal Engineering Research Center of the U.S. Army Corps of Engineers, the Coastal Research Center of Woods Hole Oceanographic Institution (WHOI), and NOAA National Sea Grant No. NA86-AA-D-SG-90, WHOI Sea Grant Project No. R/O-6. The U.S. Government is authorized to produce and distribute reprints for governmental purposes notwithstanding any copyright notation that may appear hereon. Permission was given by the Office of Chief of Engineers to publish this paper. We thank Dr. James Liu and Dr. Duncan FitzGerald for reviewing an early version of this manuscript. Woods Hole Oceanographic Institution Contribution No. 8322.

References

Aubrey, D. G., and A. G. Gaines, 1982. Rapid formation and degradation of barrier spits in areas with low rates of littoral drift. *Marine Geology,* v. 49, p. 257-278.

DeBoer, G., 1964. Spurn Head: its history and evolution. *Transactions of the Institute of British Geographers,* v. 31, p. 71-89.

Ebert, J.R., 1989. Inlet-spits and island/shoal calving: a cyclic process in the development of a flood-tidal delta, Cape Cod, Massachusetts. *Abstracts of 24th Annual Meeting, Northeastern-Section,* v. 21, No.2, p. 12.

Giese, G., 1978. Barrier beaches of Chatham, Massachusetts. Report for Town of Chatham, Massachusetts, 7 pp.

Giese, G., 1988. Cyclical behavior of the tidal inlet at Nauset Beach, Chatham, Massachusetts. In: D. G. Aubrey and L. Weishar (eds.), *Hydrodynamics and Sediment Dynamics of Tidal Inlets, Lecture Notes on Coastal and Estuarine Studies,* Springer-Verlag, p. 269-283.

Goldsmith, V., 1972. Coastal processes of a barrier island complex and adjacent ocean floor: Monomoy Island-Nauset Spit, Cape Cod, Massachusetts. Ph.D. Dissertation, University of Massachusetts, 469 pp.

Hayes, M. A., 1981. An aerial photographic investigation of barrier evolution: North Beach, Cape Cod, Massachusetts. M.S.Thesis, University of Massachusetts.

Hine, A. C., 1975. Bedform distribution and migration patterns on tidal deltas in Chatham Harbor estuary, Cape Cod, Massachusetts. In: L. E. Cronin (ed.), Geology and Engineering. *Estuarine Research,* Academic Press, Inc., p. 235-252.

Hine, A. C., 1979. Mechanisms of berm development and resulting beach growth along a barrier spit complex. Sedimentology, v. 26, p. 333-351.

Kelsey-Kennard Airviews, 1987. Chatham, Massachusetts.

Leatherman, S. P., 1979. Migration of Assateague Island, Maryland, by inlet and overwash processes. *Geology,* v. 7, p. 104-107.

McClennen, C. E., 1979. Nauset Spit: model of cyclical breaching and spit regeneration during coastal retreat. In: S. Leatherman (ed.), *Field Trip Guide Book for Eastern Section-SEPM,* p. 109-118.

Nicholls, R., 1984. The formation and stability of shingle spits. *Quaternary Newsletter,* v. 44, p. 14-21.

Ogden, J. G., 1974. Shoreline changes along the Southeastern coast of Martha's Vineyard, Massachusetts for the past 200 years. *Quaternary Research,* v. 4, p. 496-508.

Pierce, J. W., 1970. Tidal inlets and washover fans. *Journal of Geology,* v. 78, p. 230-234.

U.S. Army Corps of Engineers, 1957. Chatham, Massachusetts, Beach Erosion Control Study. 85th Congress, 1st Session, House Document, 167, 37 pp.

U.S. Army Corps of Engineers, 1968. Survey Report: Pleasant Bay, Chatham, Orleans, Harwich, Massachusetts. Dept. of Army, New England Division, Waltham, Massachusetts, 61 pp.

U.S. Army Corps of Engineers,, 1989. A study of the effects of the new breach at Chatham, Massachusetts. Dept. of the Army CERC, Reconnaissance Report, Vicksburg, Mississippi, 164 pp.

U.S. Department of Commerce, 1987, 1988, 1989. Tide Tables, East Coast of North and South America. Washington, D.C., U.S. Government Printing Office.

Weidman, C. R., 1988. Climatic factors and scarp retreat, inlet-spits and island/shoals, on a Cape Cod barrier island adjacent to a developing inlet, Chatham, Massachusetts. B.S.Thesis, State University of New York College at Oneonta, Oneonta, N.Y., 120 pp.

Weidman, C. R., and J. R. Ebert, 1988. Climatic factors and morphological change on a Cape Cod barrier island, Chatham, Massachusetts. *Abstracts of Annual Midyear Meeting-SEPM,* v. 5, Columbus, Ohio, p. 57.

Weidman, C. R., and J. R. Ebert, 1989. Dune scarp retreat as a function of meteorological and tidal controls in an area adjacent to a developing tidal inlet, Cape Cod, Massachusetts. *Abstracts of 24th Annual Meeting, Northeastern-Section,* v. 21, No. 2 , p. 74

Wood, J. F., 1976. The strategic role of perigean spring tides in nautical history and North American coastal flooding . U.S. Dept. of Commerce, NOAA, 538 pp.

7

Effects of Multiple Inlet Morphology on Tidal Exchange: Waquoit Bay, Massachusetts

David G. Aubrey, Thomas R. McSherry and Pierre P. Eliet

Abstract

Waquoit Bay, a nearly-enclosed embayment incised into outwash deposited during the last glaciation, has undergone dramatic geomorphic changes during the past century. Storms and to an extent human influence have varied the number of inlets in the Waquoit system through time, from a low of a single inlet to a high of three inlets. Water exchange rates between separate embayments within the system vary according to the number of inlets. The embayment is separated by a narrow tidal channel (Seapit River) into a western Bay (Eel River) and eastern Bay (Waquoit Bay). When only Waquoit Inlet was functioning, circulation and mixing processes in Eel River were sluggish, and residence times much longer. As the second inlet opened into Eel River, the circulation patterns changed dramatically. Residence times in Eel River were reduced, and residual currents were from Eel River into Waquoit Bay. Following Hurricane Bob in 1991, a third inlet opened, in the barrier protecting Eel River. This new inlet created increased exchange into Eel River, but its effect in general was smaller than the addition of the second inlet.

Formation and Evolution of Multiple Tidal Inlets
Coastal and Estuarine Studies, Volume 44, Pages 213-235
Copyright 1993 by the American Geophysical Union

Observations during a fifty-year period show the two-inlet system at Waquoit
to be relatively stable, in spite of some general theoretical work indicating
two-inlet systems are unstable. Waquoit is likely stable because of its
geometry, with a narrow tidal channel (Seapit River) separating two other-
wise distinct embayments. Since these embayments are shallow, nonlinear
tidal distortion is considerable, and the assumption of the theoretical stability
models are violated. Thus, shallow embayments having length scale of order
4 km can be serviced by two stable inlets, under certain conditions. Since the
range of stability conditions is not examined here, further studies are required
to clarify the stability constraints to enable more effective management of
these valuable estuarine systems.

Introduction

Geology

Estuarine coastlines make up 80 to 90% of the U.S. east coast (Friedrichs and
Aubrey, 1988). Changes in morphology of inlets and barrier beaches
separating the open sea from estuaries have resulted in efforts either to
monitor the changes or to slow them when such changes threaten habitation.
These changes occur either as a result of natural forces and self-equilibration,
or as a response to man-made alterations. Waquoit Bay, a National Estuarine
Research Reserve on the south coast of Cape Cod, Massachusetts (Figure 1),
has undergone a series of changes in response to both natural and human
influence. Important recent changes occurred in August 1991, as Hurricane
Bob swept through the area.

Waquoit Bay is incised into the outwash plain deposited from the retreating
Cape Cod Bay glacial lobe (Oldale, 1976). These outwash deposits, termed
the Mashpee Pitted Plain, are composed of well-sorted, fluvially-bedded
gravelly sand and gravel. Sub-bottom reflections show depth to bedrock at
150 m below the land surface. The loose glacial sediments are readily molded
by natural forces to produce the changes observed during the last 100 years.

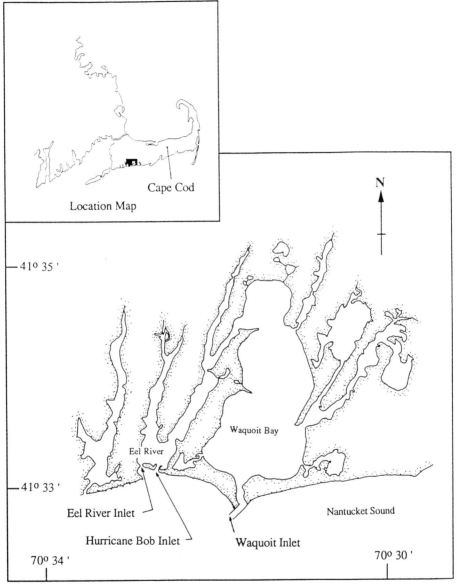

Figure 1. Location map for Waquoit Bay.

Historical Beach Evolution

Waquoit Bay is a series of interconnected drowned valleys or seepage channels forming complex subestuaries. The Bay has been connected with Nantucket Sound through several ephemeral inlets. A study (Eliet, 1990) of the historical evolution of the barrier beach provided detailed quantification of the changes through the use of aerial photographs since 1938, and navigational charts dating to 1781, to describe qualitative changes in the Bay. Before 1918 the barrier beach was continuous except for a single natural inlet into Waquoit Bay (Figure 2a). This inlet shifted its position through time in response to local wind and wave conditions. The Army Corps of Engineers (ACE) stabilized the main inlet with a jetty on its eastern side in 1918, then added a western jetty in 1937; they also installed a groin field on the outer barrier beach south of Eel River in 1937. In 1938 a hurricane created an overwash and breach into Eel River (Figure 2b), which soon developed into a new inlet. As the U.S. Army used Washburns Island during the second World War for training, the breach was filled in 1941 to support a road to the island. More groins were placed on the repaired beach facing Nantucket Sound (Figure 2c) at this time.

The ACE removed the fill in 1944, recreating the 1938 breach, but left the groins in place. This inlet changed shape in subsequent years, in response to the local waves and altered sediment supply. By 1955 the spit created by the breach had retreated towards Eel River, most likely in response to a sediment supply (Figure 2d) being reduced by man-made structures. In addition to groins and jetties, protection of adjacent upland by seawalls and revetments has further limited sediment supply. As the spit elongated it also thinned (Figure 2e, 1968); during Hurricane Bob in August 1991, storm surge left a new breach into Eel River (Figure 2f). Since the initial breach in 1938, the barrier beach has retreated 180 meters (Eliet 1990). The three inlets persisted through March, 1992, raising the question of competition between the inlets, and what changes have occurred in tidal exchange in the Bay since the inception of these recent morphological changes.

Methods

Stability Considerations

Given the cross-sectional area of each inlet and present inlet dynamics, each inlet will have a tendency to erode or accrete (Escoffier, 1940). A closure surface, used to evaluate inlet stability, is defined as the velocity/area relationship for an equilibrium inlet given a total bay surface area served by

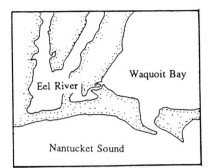

a) Before 1918

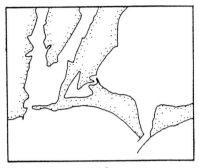

b) 1938

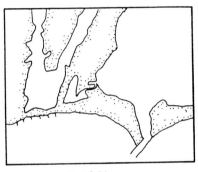

c) 1941

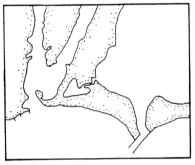

d) 1955

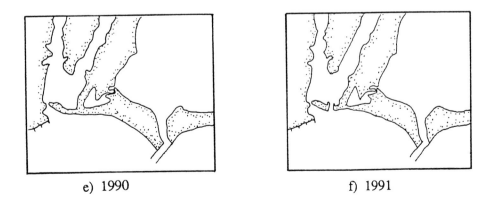

e) 1990

f) 1991

Figure 2. Inlet changes during the last century.

an inlet, the inlet width, length, friction coefficient, and entrance/exit losses. Van de Kreeke (1990) applied closure surfaces to a two-inlet embayment to show that a natural multiple-inlet system is unstable. A similar calculation is made here, where three inlets serve 6.9×10^6 m² of surface area (determined from navigational charts), with an annual mean tidal range of 0.6 meters at the Nantucket Sound entrance. The Waquoit Bay lobe has a surface area of 5.8 $\times 10^6$ square meters, and the Eel River lobe has a surface area of 1.1×10^6 square meters.

The closure velocities for each of the three inlets (for which equilibrium exists given the above parameters) and varying cross-sectional areas A_i are calculated from the following equations:

$$\left[\frac{1}{2g}\right]^2 \left[\frac{8}{3\pi}\right]^2 [A_b\sigma]^2 \, B_i^2 \hat{u}_i^4 = [A_b\eta_o\sigma]^2 - \left[\sum_{i=1}^{3} \hat{u}_i A_i\right]^2 \qquad (1)$$

$$B_i = \frac{2F_i L_i W_i}{A_i} \qquad\qquad\qquad\qquad (2)$$

where W_i, L_i, F_i are the width, length, and friction term of each of the three inlets, A_b is the total bay area, σ is the period of the tidal oscillation (12.42 hrs), and η_o is the range of the tide.

The assumption is made that the surface of the Bay fluctuates uniformly, and that each of the three inlets hydrodynamically contribute to all parts of the subestuary. Since some portions of the Bay are almost completely isolated from water entering through different inlets, and field measurements show the tidal propagation does not result in a level surface, these equations are only approximate. As an example, equation 1 for determining u_1 requires knowledge of u_2 and u_3. The values of u_2 and u_3 are taken from a one-dimensional model (Speer, et al. 1985) that predicts velocity assuming a well-mixed water column. The maximum velocity at each inlet is assigned to u_2 and u_3.

For the inlet parameters in Table 1, the closure surfaces are given in Figure 3, where the solid line is for Waquoit Inlet, the dashed line is for Eel River (1938) Inlet, and the dash-dot line is for Hurricane Bob (1991) Inlet. Present parameters are plotted as a circle for Waquoit Inlet, an asterisk for Eel River Inlet, and a plus sign for Hurricane Bob Inlet. Assuming that Waquoit Inlet

Table 1. Inlet Parameters (Width, length, and area from navigational chart).

Inlet	Waquoit Inlet	Eel River Inlet	Hurr. Bob Inlet
Width	74 m	33 m	95 m
Length	762 m	33 m	66 m
Area	203 m^2	88 m^2	33 m^2
MaxVelocity (model)	0.75 m/sec	0.51 m/sec	0.10 m/sec

is presently in equilibrium (suggested by its longevity), the equilibrium bay surface A_b in this analysis is set to 4.2 x 10^6 m^2 rather than the actual total bay area of 6.9 x 10^6 m^2. This reduced value causes the closure surface (solid line) and circle (Waquoit area and maximum velocity) to intersect. This configuration is stable since a decrease in cross-sectional area would increase maximum velocity which would slow the shoaling. Using the same surface area of 4.2 x 10^6 m^2, the Eel River (1938) closure surface (dashed line) does not intersect the present inlet geometry (*). If the surface area were decreased to 3.8 x 10^6 m^2, the curve and point do intersect, however, demonstrating the sensitivity of the closure curve. Since it has persisted for half a decade, this western inlet appears to be stable. Clearly the Bay area that Waquoit Inlet services is not the same as that for Eel River Inlet. Using the smaller area, the intersection is on the stable side of the critical area, so a decrease in area causes an increase in velocity. Eel River and Hurricane Bob inlets service the same Bay surface area. Applying the 3.8 x 10^6 m^2 area still leaves a discrepancy in present conditions (+) and equilibrium conditions (dash-dot line) as in Figure 3. The flow and area are so small that reaching an equilibrium condition is not likely. The expected future for Hurricane Bob (1991) Inlet is closure if there is sufficient sediment transport capacity in the system. This may take time since the sediment supply has been decreased from both directions.

Conclusions can be drawn from these calculations. First, in a complicated embayment such as Waquoit Bay, the two primary inlets interact essentially with different portions of the embayment. Stability calculations cannot be based on equal surface areas. Instead, complicated tidal propagation and dissipation permit some hydrodynamic decoupling within the embayment. Since the linear Escoffier model does not permit us to calculate this decoupling, we use the approximate equations to determine the equilibrium surface areas, since the two inlets have been stable. The total volume for stability is

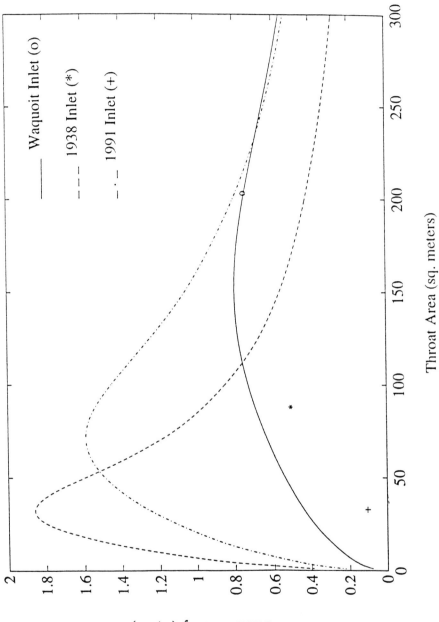

Figure 3. Equilibrium flow curves for three inlets.

8×10^6 m² (4.2×10^6 m² for Waquoit Inlet and 3.8×10^6 m² for Eel River Inlet), about 16% greater than the measured surface area. Upstream decrease in tidal range will require more area to provide the same tidal prism; hence this 16% increase in effective area may reflect nonlinear tidal dissipation and propagation within the system, as well as tidal residuals between the basins. Finally, in no case is the three-inlet system stable.

Numerical Model

A one-dimensional diagnostic model that determines sea-surface, discharge, and concentration in Waquoit Bay was applied to study the hydrodynamic effects of changing morphology. The model grid is a linked series of branches covering each main subestuary, resolving the estuary into two-dimensional cross-sections (Figure 4a). The hydrodynamics for the model are described in Speer et al. (1985); here we describe only the mixing routine based upon the continuity equation for mass:

$$\frac{\partial(AC)}{\partial t} + \frac{\partial(QC)}{\partial x} = K_s \frac{\partial}{\partial x}\left[A\frac{\partial C}{\partial x}\right] \tag{3}$$

where A is the area through which mass is transported, Q is the volume flux of water through this area, K_s is the longitudinal dispersion coefficient, and C is the concentration. The differencing is central in space and forward in time:

$$\frac{\partial(AC)}{\partial t} = \frac{\left(A_j^{n+1} C_j^{n+1} - A_j^n C_j^n\right)}{\Delta T} \tag{4}$$

$$\frac{\partial(QC)}{\partial x} = \frac{\left(Q_j^n\left[C_{j+1}^n - C_{j-1}^n\right] + C_j^n\left[Q_{j+1}^n - Q_{j-1}^n\right]\right)}{2\Delta X} \tag{5}$$

$$K_s\frac{\partial}{\partial x}\left(A\frac{\partial C}{\partial x}\right) = \frac{K_s A_j^n}{\Delta X^2}\left(C_{j+1}^n - 2C_j^n + C_{j-1}^n\right) \tag{6}$$

$$K_s \frac{\partial}{\partial x}\left(A\frac{\partial C}{\partial x} \right) = \frac{K_s}{2\Delta X^2}$$

$$\left(\left[A_{j+1}^n - A_j^n \right]\left[C_{j+1}^n - C_j^n \right] + \left[A_j^n - A_{j-1}^n \right]\left[C_j^n - C_{j-1}^n \right] \right) \tag{7}$$

where ΔX is the discrete spacing between nodes, ΔT is the time increment, the subscript j denotes the node number, and the superscript n is the time indexing.

The space differences are performed so as not to exclude information from the grid center. This reduces error when cross-sectional areas change rapidly. Error from this differencing results from large gradients in concentration or area resulting in negative concentrations at the next time step. This problem is usually confined to startup conditions, and goes away as the solution becomes more smooth. It can be further reduced by adding tapers to the initial concentrations. The node- adjacent junctions connecting several branches, or next to river heads, require a different treatment.

For node-adjacent junctions, consider Figure 4b and the areas at nodes (2,1) and (3,1). These are summed to equal A1, and the areas at (2,2) and (3,2) are summed to equal A2. Mass at these nodes are also lumped together, then divided by the total areas A1 and A2. Volume fluxes are also collected. Two fictitious nodes based upon the properties at the first two nodes in the exiting branches from the junction are created. Denoting these nodes as (F,1) and (F,2), the model then calculates the concentration at (1,N) using the same differencing as above (eq. 2-5) and nodes (1,N-1), (1,N), and (F,1) from the previous time step. Likewise, the concentration at node (F,1) is based upon information at nodes (1,N), (F,1), and (F,2) from the previous time step. It can be shown that the new concentration at (2,1) and (3,1) is the new concentration at (F,1). Mass is neither created nor lost using this technique.

At branch heads mass cannot be advected through the wall. The hydrodynamic condition of Q equalling zero satisfies this criterion. Neither can mass be dispersed through the wall, represented as $K_s (\partial C/\partial x) = 0$. Equating A and C of a fictitious upstream node to the values at the head of a branch, the same central differencing as above can be employed without loss or creation of mass.

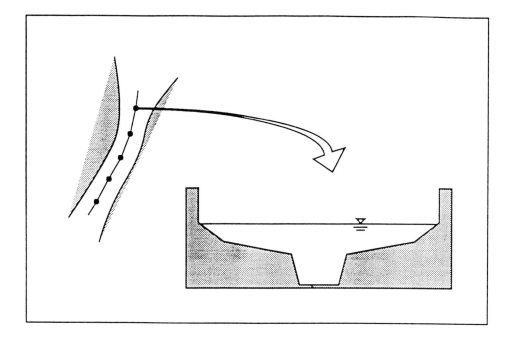

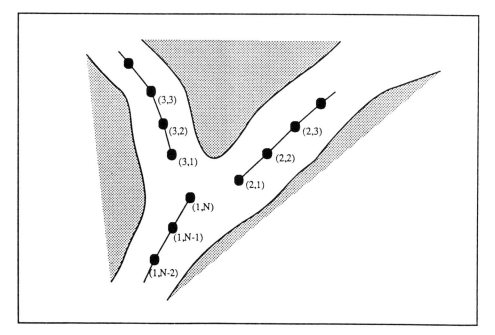

Figure 4. a) model cross-sectional geometry; b) grid junction treatment.

Verification of Numerical Model

Results (Jaworski and Clark, 1970) of a field study from the Potomac River estuary (Figure 5) are used to verify the model. The semi-diurnal tidal amplitude at the mouth is 0.215 meters. For a friction factor of 0.0017, the range at the head from the model is 0.83 meters with a phase lag between the mouth and the head of 6 hours 41 minutes. The measured range at the head was 0.88 meters with a phase lag of 6 hours 53 minutes. Field salinity at the mouth was measured at 17.2 ppt, and salinity at upstream stations were recorded for 39 tidal cycles. Figure 6 shows salinity comparisons after 1200 hours of model runtime for a K_s of 200 m²/sec at each station, and freshwater input at the head of 112 m³/second. Table 2 gives station, line number, and symbol for the figure.

Table 2. I.D. for Figure 6.

Station (Figure 6)	Line (from top)	Symbol (field data)
Piney Pt.	1	o
Kingcoscipo Pt.	2	*
Wicomico River	3	⊕
U.S. 301 Bridge	4	+
Smith Pt.	5	o

Agreement is generally good between model and field results. Differences are most likely due to variable freshwater inputs after the 15th tidal cycle. The match improves nearer the mouth of the estuary. All field data exhibit a freshening starting after the 15th tidal cycle at the Smith Point station and continuing with a time lag downriver, suggesting a storm during the middle of the field acquisition period.

Results

Flushing

One residence time can be defined as the bay volume divided by the gross inflow during one half tidal cycle. Using a mean tidal range of 0.6 meters, and

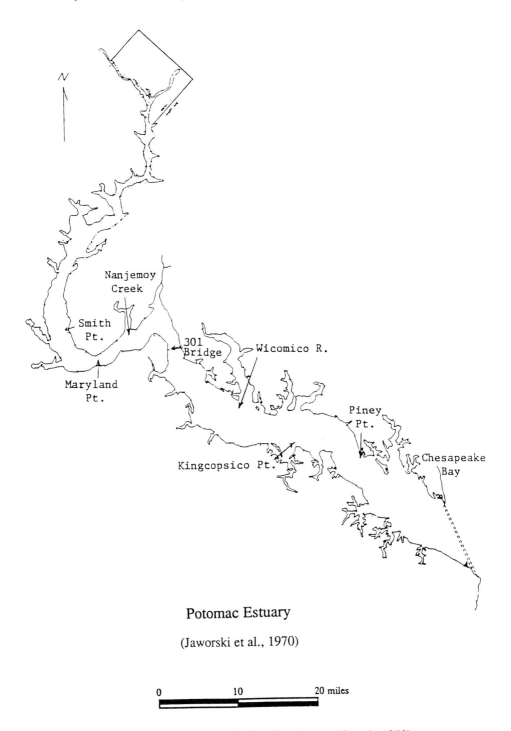

Potomac Estuary

(Jaworski et al., 1970)

0   10   20 miles

Figure 5. Potomac River estuary (taken from Jaworski et al., 1970).

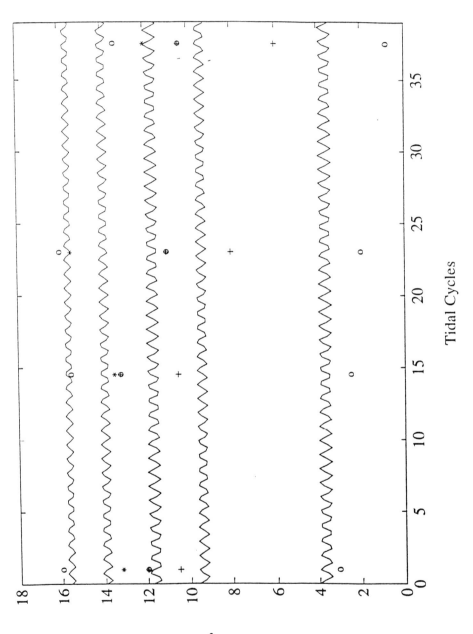

Figure 6. Salinity comparisons along Potomac estuary (see Table 2 for symbol I.D.).

bay mean water volume of 7.5×10^6 m^3, residence time is calculated for three conditions: a) Single inlet (Waquoit); b) two inlets (Waquoit and Eel River [1938] inlet); and c) three inlets (Waquoit, Eel River [1938], and Hurricane Bob [1991] inlets). Residence time for condition (a) is 34.6 hours. For (b) the time is 32.8 hours, and for (c) the time is 32.4 hours. This estimate assumes complete exchange between water near the inlet and water in the subestuaries over only two-and-a-half tidal cycles. Inclusion of more complete mixing physics would give a better estimate of residence time. Defining residence time as the time it takes for 95% of some initial material weight to leave the system (T_{95}), and seeding the entire Bay with one unit of concentration, an exponential decay is developed for each of the three scenarios using a dispersion coefficient (K_s) = 30 m^2/sec. The decay is calculated for a given period of time. Figure 7 shows the model results, for a decay curve (dashed line) of the form:

$$Y = \alpha e^{\beta t} \tag{8}$$

where Y is the percent mass remaining in the system, and t is time. The T_{95} threshold can be defined as:

$$T_{95} = \frac{1}{\beta} \ln\left(\frac{0.05}{\alpha}\right) \tag{9}$$

For the three conditions, (a-c), the following coefficients and T_{95} residence times have been calculated:

Table 3. Decay coefficients and T_{95}

Inlet Scenario	α	β (hrs^{-1})	T95 (hrs)
a	0.96707	-0.01421	208.5
b	0.93901	-0.01683	174.3
c	0.95485	-0.01684	175.2

The T_{95} times for conditions (b) and (c) conditions are similar, which shows inability of an additional inlet to change on the flushing characteristics of the Bay. Dispersion processes yield a clearer picture of water movement within the system than using tidal prism estimates alone. Water which starts in the upper reaches of the Bay has a longer residence time than water near the inlets.

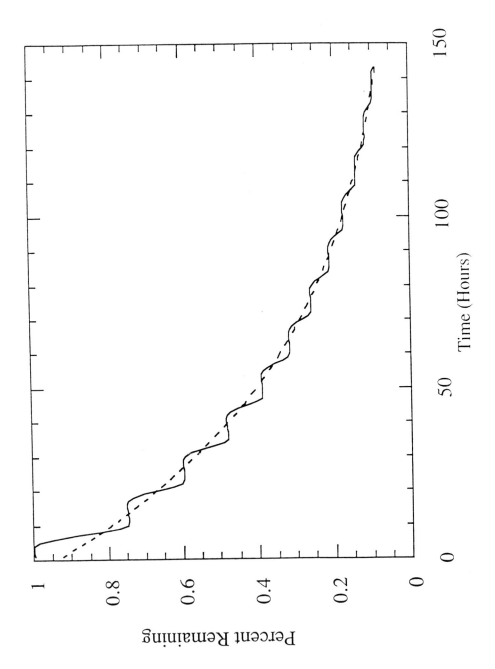

Figure 7. Mass decay curve in Waquoit Bay. Dashed line is exponential fit.

This near-inlet water efficiently exits the system and is replaced several times over before remote waters find their way to the inlets. The T_{95} times decrease, of course, as freshwater input increases. The numbers in Table 3 are without freshwater input.

Water Exchange

Water exchange rates differ for the three inlet scenarios, where exchange coefficients are now determined by splitting the Bay into 22 subelements (Figure 8) using the dispersion model to trace water parcels. After two tidal cycles, tracer mass is compared to the original distribution in amount and location. This information is condensed and used in a coarse time step ($\Delta T = 24.82$ hrs) model which moves water and particles according to the calculated exchange coefficients. This step is helpful for biologists who are interested in larger scale changes both spatially and temporally than what is studied in hydrodynamics. The large time step means much smaller-capacity computers can apply hydrodynamic results over yearly time scales. In this application, 1 kg/m³ concentration tracer is released in box 19 for each of the three scenarios, then tracked for eight days. The differences in the evolution of the mass movement between (a) and (b) are demonstrated in Figure 9. Immediately after startup and continuing for the duration of the run, there is a considerable difference in the mass (16000 kg) left in boxes 19 through 22 for scenarios (a) versus (b). The two-inlet system transports water from the area through the inlets quickly, suggested by the positive peak of mass in box 10 on the second day which fades away on the third, out through the inlet. Mass does not enter boxes 10 through 12 for the one-inlet case until the fourth day, shown as the large negative area of 4000 kg. The zero-difference region over boxes 1 through 8 suggests little difference in water exchange between (a) and (b). Figure 10 shows the difference between (b) and (c). The differences are smaller than between (a) and (b), but suggest for three inlets a transport of mass into boxes 15 through 18, as well as less efficient transport into boxes 10 through 12 in the first four days. After four days, there is a net decrease in mass for (c) compared to (b). This figure suggests greater decoupling between the western and eastern sides of the Bay for three inlets, but only slightly.

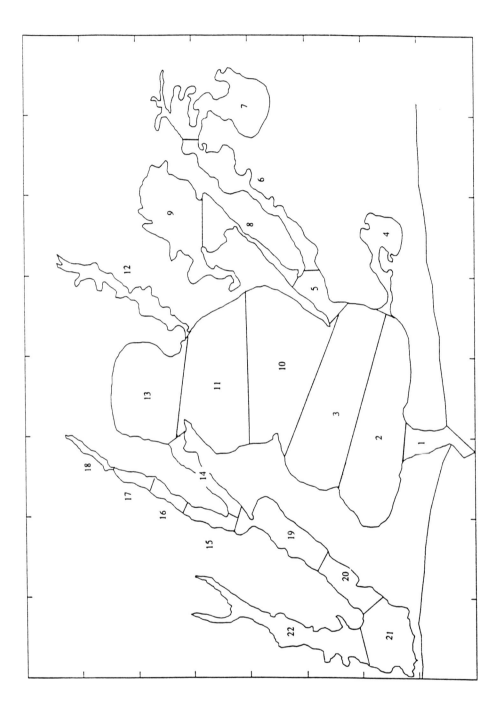

Figure 8. Sub-elements of water parcels in Waquoit Bay.

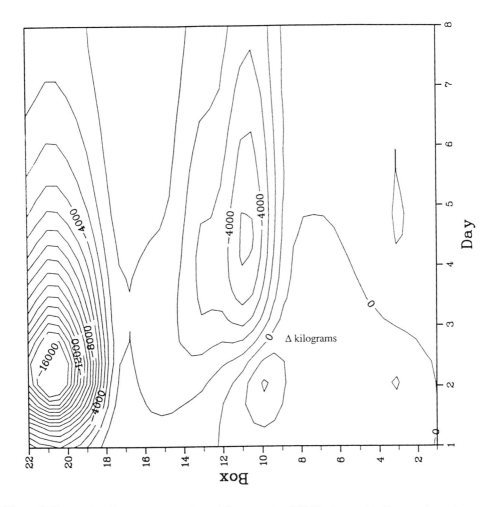

Figure 9. Comparison between one- and two-inlet scenarios (1000 kg intervals). Box number refers to designated sub-element of Figure 8. Day corresponds to time after the initial release of the substance. The contours show mass differences between these boxes for each scenario. For example, on Day 4.5, there is roughly 5000 kilograms more material in Box 11 if only Waquoit Inlet is operational versus the addition of Eel River Inlet.

Residual Transport

One-dimensional tidal residual transport illustrates the pathways for net dispersal of materials from the western embayment to the eastern embayment. Averages taken over several tidal cycles demonstrate the net transport of material from one system to the next. Numerical experiments were performed

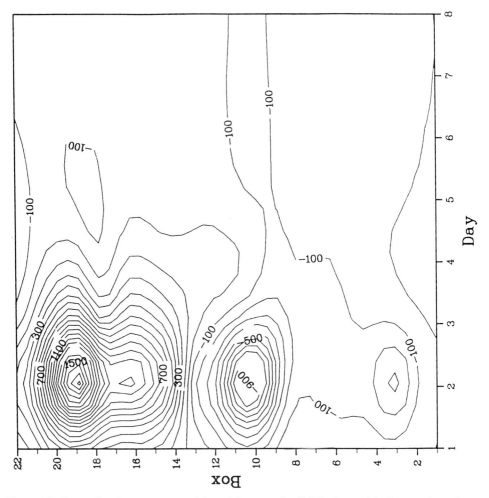

Figure 10. Comparison between two- and three-inlet scenarios (100 kg intervals). Box number refers to designated sub-element of Figure 8. Day corresponds to time after the initial release of the substance. The contours show mass differences between these boxes for each scenario. For example, on Day 2, there is roughly 1700 kilograms less material in Box 19 if only Waquoit Inlet and Eel River are operational versus the addition of Hurricane Bob Inlet.

for two scenarios: zero freshwater input and mean freshwater input into Childs River of 9500 m³/day.

For the case of a single inlet (Waquoit Inlet) with no freshwater inflow, there is by definition no residual transport between the two embayments. However, once freshwater inflow is added, Eel River embayment is a source of tracer to Waquoit Bay itself. For the two inlet system, the situation changes. The residual water transport is from Eel River to Waquoit Bay at a rate of 3725 m³

per day. This increases to 7125 m³ per day from the Eel River side to the
Waquoit Bay side when freshwater is added to Childs River. Flow through
Seapit River changed phase and sense of symmetry following the appearance
of Eel River Inlet (Figure 11). The velocity phase after the second inlet was
introduced changed by 180°, and the flow from Eel River into Waquoit Bay
became noticeably asymmetric. When the model is run with Waquoit, Eel
River, and Hurricane Bob inlets, the residual flow is 3735 m³ per day, similar
to the two inlet case. When freshwater is included, this increases to 7135 m³
per day. Thus, the effects of Hurricane Bob Inlet are negligible compared to
the effects of Eel River Inlet.

Conclusions

By applying stability methods to the different three inlet configurations that
have existed in Waquoit Bay during the last century, and considering the
effects of man-made structures, the two-inlet condition appears to be stable.
Hurricane Bob (1991) Inlet currently is too small to maintain itself, although
the inlet may support increased flow to the detriment of flow through the Eel
River (1938) Inlet. The inlet may even migrate westward to connect to the Eel
River inlet. The barrier beach has retreated significantly to the north, as a
result of decreased sediment supply; this barrier migration may ultimately
determine the fate of the two inlets. Absent this rapid barrier migration, the
two inlets serving Eel River may persist because of this reduced sediment
transport rate. At Popponesset Spit, a barrier beach located five km to the east,
sediment transport conditions similarly are low (Aubrey and Gaines, 1983).
A temporary breach remained open for half a decade there, as insufficient
longshore transport was available to close it. With limited longshore
transport, Hurricane Bob (1991) Inlet may also persist.

The Bay has improved its flushing characteristics as more inlets were added
to the system. Although Eel River (1938) Inlet arose from largely non-natural
modifications, it now allows water in the western lobe of the system to be
renewed more quickly with water from other parts of the system and from
Nantucket Sound. Water exchange varied only slightly after the third inlet
was formed in 1991, but residual flow patterns suggest a more direct
communication of water in upper reaches of western lobe rivers with water

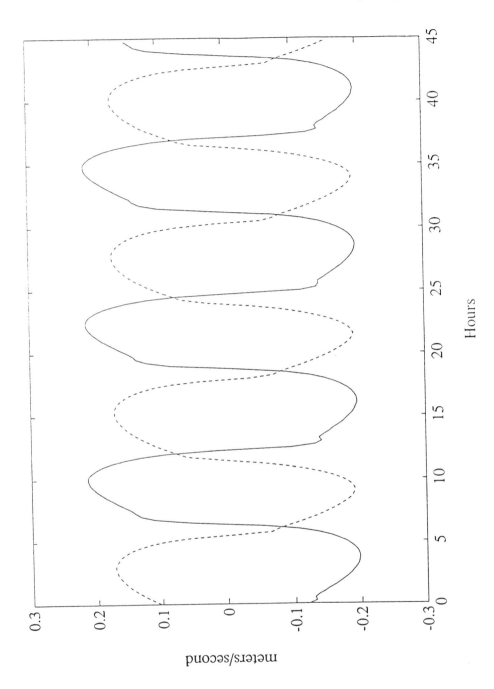

Figure 11. Velocities in Seapit River. Solid line indicates case of only Waquoit Inlet. Dashed line indicates addition of Eel River Inlet. With the addition there occurs an almost 180° phase shift, as well as a more assymetric shape.

from the 1938 and 1991 inlets. This factor may affect land-use planning in the area, particularly as existing inlets close or new inlets are formed. The models and techniques shown here prove to be flexible tools to investigate long-term evolution of embayments given sudden alterations in geometry. The nature of estuaries separated from ocean waters by thin, moveable barriers lend themselves to such study.

Acknowledgments

This work was supported by a subcontract from Boston University to WHOI, National Science Foundation LMER program OCE-89-14729. Wayne Spencer helped with the field work, and Pamela Barrows finalized the manuscript. Woods Hole Oceanographic Institution Contribution No. 8323.

References

Aubrey, D.G. and A.G. Gaines, Jr. 1982. Recent Evolution of an active barrier beach complex: Popponesset Beach, Cape Cod, Massachusetts. WHOI Tech. Rept. 82-3, 77 pp.

Bruun, P. 1978. *Stability of tidal inlets*. Elsevier Scientific Publishing Co., 510 pp.

Eliet, P.P. 1990. An historic analysis of the changing geomorphology of the Waquoit Bay estuarine system. WHOI summer fellowship student rept., 12 pp.

Escoffier, F.F. 1940. The stability of tidal inlets. *Shore and Beach*, v. 8, p. 114-115.

Friedrichs, C.T. and D.G. Aubrey 1988. Non-linear tidal distortion in shallow well-mixed estuaries: a synthesis. *Estuarine, Coastal and Shelf Science*, v. 27, p. 521-545.

Jaworski, N.A. and L.J. Clark 1970. Physical Data Potomac River Tidal System Including Mathematical Model Segmentation. Technical Report No. 43, Chesapeake Technical Support Laboratory, Federal Water Quality Administration.

Oldale, R.N. 1976. Notes on the generalized geologic map of Cape Cod. USGS, Open file report 76-765, 23 pp.

Speer, P.E. and D.G. Aubrey 1985. A study of non-linear tidal propagation in shallow inlet/ estuarine systems. Part II: Theory. *Estuarine, Coastal and Shelf Science*, v. 21, p. 207-224.

U.S. Army Corps of Engineers, 1962. Beach erosion control report on cooperative study of Falmouth, Massachusetts. U.S. Army Engineer Division, New England, 32 pp.

van de Kreeke, J. 1990. Can multiple tidal inlets be stable? *Estuarine, Coastal and Shelf Science*, v. 30, p. 261-273.

List of Contributors

David G. Aubrey
Department of Geology and
Geophysics
Woods Hole Oceanographic Institution
Woods Hole, MA 02543

James R. Ebert
Department of Earth Sciences
State University of New York
College at Oneonta
Oneonta, NY 13820-4015

Pierre P. Eliet
St. Margarets
Castlepark Road
Sandycove
Co. Dublin, Ireland

Duncan M. FitzGerald
Coastal Environmental Research Group
Geology Department
Boston University
Boston, MA 02215

Carl T. Friedrichs
Department of Geology and
Geophysics
Woods Hole Oceanographic Institution
Woods Hole, MA 02543

Graham S. Giese
Department of Geology and
Geophysics
Woods Hole Oceanographic Institution
Woods Hole, MA 02543

James T. Liu
Institute of Marine Geology
National Sun Yat-Sen University
Kaohsiung, Tiwan 80424

Thomas R. McSherry
Department of Geology and
Geophysics
Woods Hole Oceanographic Institution
Woods Hole, MA 02543

Todd M. Motello
Coastal Environmental Research Group
Geology Department
Boston University
Boston, MA 02215

Paul E. Speer
Center for Naval Analyses
Alexandria, VA 22302-0268

Donald K. Stauble
Coastal Engineering Research Group
Waterway Experiment Station
U.S. Army Corps of Engineers
P.O. Box 631
Vicksburg, MS 39181

Christopher R. Weidman
Joint Program MIT/WHOI
Department of Geology and
Geophysics
Woods Hole Oceanographic Institution
Woods Hole, MA 02543